AGRICULTURAL GEOGRAPHY

John R. Tarrant

Previous studies of agricultural geography either present a descriptive account of the world's major agricultural systems or provide detailed analyses of the agriculture of a single country or region. Neither of these approaches reflect recent trends in geography towards hypothesis testing and model building and the development of more sophisticated methods of analysis.

Dr Tarrant's study breaks new ground by discussing the value of models in the study of agriculture and explaining new analysis methods, pointing out their weaknesses and strengths, without assuming a great knowledge of quantitative techniques. The problems of applying powerful methods of analysis to inadequate data are put in perspective by a major section of the book which reviews the availability and suitability of many types of agricultural data with examples from the United Kingdom, the United States, Sweden and elsewhere.

Taking account of recent changes in the size of the European Economic Community, and continuing problems of excess food in many parts of the world, an important chapter is devoted to agricultural marketing, concentrating especially on the role of governments in attempts at stabilisation of agricultural production, distribution and marketing. The loss of rural agricultural land to urban uses is of equal concern in many developed countries and the processes of this change are examined, placing emphasis on the impact of such demands for land on agriculture.

In setting down a methodological framework for the study of agricultural geography and in presenting an analysis of two aspects of agricultural change which have proved difficult to incorporate in agricultural location models, Dr Tarrant's text provides a basic dimension for all courses in agricultural geography.

Dr John R. Tarrant is lecturer in Environmental Science at the University of East Anglia, and in 1973 was visiting lecturer in Geography at the University of Canterbury, Christchurch, New Zealand.

Jacket illustration: Courtesy NASA/Goddard Space Flight Centre.

PROBLEMS IN MODERN GEOGRAPHY

Series Editor — Richard Lawton *Professor of Geography, University of Liverpool*

PUBLISHED

J. Allan Patmore	*Land and Leisure*
David Herbert	*Urban Geography: A Social Perspective*
Kenneth Warren	*Mineral Resources*
John R. Tarrant	*Agricultural Geography*

IN PREPARATION

Derelict Land: Origins and Prospects of a Land-use Problem	Kenneth Wallwork
New Communities	H. Brian Rodgers
World Automotive Industry	G. T. Bloomfield
The Rebuilding of Europe: A Geographical Analysis	Mark Blacksell
Recreational Geography	Patrick Lavery (editor)
Water Supply	Judith Rees
Political Geography of the Oceans	V. R. Prescott
Nuclear Power and the Energy Revolution	P. R. Mounfield

PROBLEMS IN MODERN GEOGRAPHY

JOHN R. TARRANT

Agricultural Geography

A HALSTED PRESS BOOK

JOHN WILEY & SONS
New York

Published in the U.S.A. by
Halsted Press, a Division of
John Wiley & Sons, Inc. New York

Library of Congress Cataloging in Publication Data

Tarrant, John Rex.
Agricultural geography.

(Problems in modern geography)
"A Halsted Press book."
Includes bibliographical references.
1. Agricultural geography. I. Title.
S495.T35 630 73-11883
ISBN 0-470-84565-1

Printed in Great Britain

Contents

List of Figures

List of Tables

CHAPTER ONE

Introduction

THE description of farming as a 'way of life' belies the strong economic forces which underlie most forms of agricultural production. Farming, whether for a cash income, for subsistence or both, is primarily a way of economically supporting a farmer and his family, which implies that the goods he is producing must be either used by him or sold or exchanged in markets. This does not mean that economic principles are always uppermost but that the impact of the constantly changing patterns of demand, supply, price, government support policy and subsidy structure are considerable and, at least in the developed economies, almost universal. On the other hand there is a wide variety of possible individual reactions to any economic structure. In Britain, for example, these range from the sheep-based hill farmer of central Wales, having very little choice but to continue much as his father and grandfather had done before him, to the 'hobby farmer' of East Anglia whose involvement in agriculture may have an economic reason in his personal taxation situation and may be more a result of interest in the land or the game that it can produce than in any desire or financial need to produce agricultural products for market. The economic facts of agricultural life never act in an entirely deterministic way but rather set limits within which farmers are able to operate; they define the freedom of choice.

The other major control on farmers' actions is the physical environment which again does not act in a deterministic manner but rather lays down broad controls over what can and cannot be grown and over yields which can be expected for a given input of labour, fertilisers and other factors of production. Economic and physical factors interact with each other in a complex manner and, together with certain characteristics of the farmer himself, establish a range of possible forms of production one or more of which may be chosen. To some extent the range of choice offered by the physical environment can be extended by factors of the economic environment. Investment, in the form of private capital or government assistance, can be substituted for less than ideal environmental conditions. Poor soil may be remedied by the application of fertilisers, poor drainage by new field drainage systems or poor beef fattening performance on indifferent grazing ground by intensive rearing in controlled indoor environments. Similarly a shortage of capital will be less restrictive of a farmer's freedom of action in areas where the physical environment is favourable. Greater restriction in one direction may be alleviated by greater freedom in another. In such circumstances it is in the areas of 'marginal farming' that the least freedom of action is found. Poor environmental conditions imply a limited range of options and this is aggravated by shortage or virtual non-existence of capital for investment. Co-operative organisation and government assistance may be available in such areas attempting between them to improve economic flexibility and thereby helping to overcome the limiting physical conditions.

With such interaction between the physical environment and economic circumstances a great diversity of agriculture can be expected, especially as each producer occupies only a small piece of land and, whereas physical circumstances can be more or less uniform over a large number of producers, economic circumstances are unlikely to be. The recognition of a set of economic and physical controls implies that each farmer has a 'best' solution to the problems of what agricultural products he should produce, given his environmental and economic circumstances: for several reasons this is either not the case or is of little relevance in the real world. Any one crop is rarely grown to the total exclusion of all others; not only are crops rotated to preserve soil fertility and structure but also it is hardly wise for a farmer to risk all his efforts on a single crop. A year's income may be derived from the crop and crop failure or a poor market price would lead to financial disaster. On any one farm combinations of different enterprises are found, to a greater or lesser extent dependent on one another. The physical environment may be suitable for one crop in the combination but is unlikely to be equally suited to all so that there is no simple relationship with the physical environment that determines agriculture. Kale may be produced on a farm to feed a dairy herd or barley may be grown to feed store cattle; alternatively, if barley is of high enough quality, it can be sold as a cash crop for seed or for malting and the income used to buy in feed for the store animals. There are many such combinations of farming enterprises which can be chosen by the individual farmer to suit his particular circumstances and his preferences.

Such patterns are made more complex when we consider the farmer's personal characteristics, particularly his knowledge of new developments in farming which, together with his attitudes to risk, will further restrict his freedom of manœuvre. Knowledge about the latest farming technology is not universally available and even if it were it would not be universally accepted. Evidence of the spread of agricultural innovations has shown that reactions among farmers to new ideas vary considerably and it may be many years before a new idea, even of very obvious worth, is adopted by all the farming community. Certain combinations of enterprises carry with them greater financial risks which may be acceptable to some farmers because of the prospect of greater economic return. In other cases, perhaps where the farmer has a large family to support and little capital to fall back on in the event of failure, such risks may be totally unacceptable. The attitude of tenant farmers and owner occupiers will be different, the one using the land to make a living, the other owning the land and hoping to pass it on to his children. Although there is, theoretically, a 'best' agricultural system for any portion of the earth with any given state of the market and price structure, because of the very wide range of choice and the wide range of individuals making this choice it is unlikely that this optimum solution is found in more than a very small minority of cases.

A further complication is introduced when we consider that agriculture is a dynamic process and, although aspects of the physical environment can be taken as stable over considerable periods of time, that the economic environment may be very unstable. Changes in demand, prices and government assistance are constantly occurring: examination of farming at any particular moment will show that there is a lag between the state of the economic and physical environment and farmers' reactions to it. Changes are constantly under way; farms become more mechanised, labour forces are reduced, hedges are removed to make larger, more economical, fields, new methods of field drainage are introduced, new strains and varieties of crops become available and new types of market are exploited. In addition to differing reactions to their physical and economic circumstances farmers also react differently to change. The ability to adopt innovations in agriculture varies very considerably from farmer to farmer; this ability is closely linked with other economic circumstances, particularly the availability of capital. In any area agriculture reflects a whole gradation of adjustment to constantly changing circumstances. No location theory for agriculture as a whole can hope to be successful if it is based on the concept of agricultural production ultimately reaching a static equilibrium where a given environment has an optimum solution independent of time.

Thus it is hardly surprising that there is such variety in agricultural practice, even over very small distances. The agricultural geographer attempts to recognise these variations at various levels, international to local, and to attempt to explain them: he must try to find general laws which apply to the spatially distributed phenomena he is investigating. William Bunge[1] has laid considerable stress on the differentiation within geography of the unique from the general and concludes that, although all features of the earth including farms appear unique in many respects, none the less they have some features in common. Farmers most certainly do not react in the same way to the pressures of the environment, but the fact that they do react provides a basis for generalisation. Although each farm, taken as an individual, can be examined to see how it is different from every other, this type of approach to the study of agriculture builds up an accumulation of information which it is hoped may eventually be brought together in a grand synthesis to provide a definitive agricultural geography. But the more comprehensive this approach, the more data are collected and the less practicable such a synthesis becomes. To approach the problem from the other direction by examining what farms are likely to have in common leads to the establishment of hypotheses and, from these, models which can be tested by empirical study. It is this approach which forms the basis of this book.

If agricultural geography is to explain the variations amongst farmers' reactions to a variety of physical, social and economic constraints, it must be done, not by simply describing these differences, but by seeking principles and, perhaps, laws which will account for them. In practice this ideal has so far remained unattainable because of the extreme complexity of farming situations and because of the limiting nature of the available information with which to establish and test such models.

As the basic unit of agriculture, the farm should be the basic unit for the study of agricultural geography. But, with approximately 150,000 farms in England and Wales alone, study based on the farm unit presents formidable problems in terms of data availability and sampling. The basic unit of agriculture is too small in relation to the total area of farming in the world. This size differential between farms and the total agricultural sector of the economy makes agricultural marketing difficult to organise and operate efficiently while dealing fairly with individual producers. Marketing boards, producers' co-operatives and policies of price control have been tried in attempts to improve the position of producers in the market place and these will be examined more fully in Chapter 7. Chapter 3 deals with problems of accumulating data about individual farms and the alternatives available. Undoubtedly the type of data available has had a profound impact on agricultural geography. If data for individual farms could be collected other than by laboriously interviewing farmers, we would have a greater understanding of spatial variations in agriculture than we now do. Not only are data about the current state of farming very difficult to obtain but in a highly dynamic situation with changes of various magnitudes taking place over different time intervals, such data are of limited value for a particular moment in time. Information on reactions to change and the process by which change is brought about in the farming community is often lacking. Until this situation improves, the lack of a dynamic explanatory model of agricultural variability will remain. Chapter 2 will present some attempted explanatory models of agriculture but, in common with many such techniques of analysis, their basic shortcoming is that of an essentially static approach to a dynamic situation.

In agricultural geography much attention is paid to the nature of spatial variations in agricultural production. In dealing with such variety it is natural that we should start by thinking in terms of methods of classification. This process is one in which we no longer regard each individual as unique but as forming part of a group with some feature or features in common: the common features now become the centre of interest. Thus in Chapter 4, dealing with various farm classifications, a fairly clear idea can be given of a group of dairy farms: although some may grow their own feed while some import it, some may be large while some are small, such differences are of relatively minor importance in the face of the common major characteristic, that of concentrating on the production of fresh milk for market. Once such types of farming are isolated they can be used as the basis for discussion about the nature of farming. As all farms exist at particular points on the earth it is possible to link together those which are spatially related, as well as of similar characteristics, to produce type of farming regions within each of which the character of agriculture is sufficiently distinctive to set it apart from other, neighbouring, regions. Spatial variation operates at all scales: there are variations within the physical environment of a single field which may lead to variations in crop yield at this scale; there may even be different crops planted in the same field. At the larger scale of the individual farm there will be more obvious differences and neighbouring farms in the same physical and economic circumstances may show considerable variation. Viewed in a larger spatial context, such inter-farm differences may become subordinate to a large-scale regional similarity: individual variation will take on the characteristic of 'noise' within a clear 'signal' of the overall similarities in the area. With an even wider view it is possible to see the agriculture of a country as a whole as being different from or similar to agriculture in other countries. For example large parts of the world practise some form of subsistence agriculture. Although based on a wide range of crops, the similarity of the economic systems makes for a single categorisation. With such continuous variation at all scales boundaries are difficult to draw and in fact the closer one examines any particular regional boundary the less distinct it appears; just as a line on a printed page viewed at a normal distance of about 40 cms appears distinct, while from closer positions under magnification the edges of the line become blurred where the printing ink has spread into the fibres of the paper. While it is obviously important to be clear about the nature of the variations we are trying to explain in agricultural geography, possibly too much attention has been paid to the definition and description of agricultural regions to the exclusion both of variations at different scales and of the definition of general principles governing these variations.

Other methods of analysis which do not include regionalisation are examined in this book and, whereas many provide only alternative methods of describing agricultural patterns, they may be able in the future to provide the bases of appropriate dynamic models which will enable us better to explain the distribution of agricultural types.

Such models will have to consider the wide range of marketing practices for different crops in different countries. Early work on agricultural location considered the role of the market in relation to its control on demand and in terms of the cost of access to it for farm produce. As a high proportion of the world's agricultural production is sold through markets, governments, in attempting to provide security and stability in their country's agriculture and often to secure a high proportion of home-based food production, use the market as the medium through which they can provide the necessary control. The importance of marketing and the position of the producer *vis-à-vis* the consumer is very considerable and is taken up in Chapter 7. With talk of environmental crisis and mass starvation it is important to remember that the main problem, at least in the short term, is not so much the production of food but its equitable marketing and distribution.

There is further concern in the developed world with the changing role of land from an agricultural resource to an urban resource. Instead of acting as a raw material in the production of crops and livestock, more and more land is used to provide space for houses, gardens, recreation, water supply, transport and many other uses. Most of these uses are totally incompatible with agriculture and the change of use is virtually irreversible. With the worldwide spread of urbanisation it is appropriate that we conclude this book with a consideration of the process of land conversion from rural to urban uses and with suggesting ways in which such a process may develop. The reduction in the stock of rural land which results from continued urban growth can be viewed not only as a loss of land to agriculture but also as a loss to recreation, wildlife and perhaps to amenity in general.

Such matters are primarily the concern of the developed nations of the world. Most under-developed nations are more concerned with the current problems of feeding a rapidly increasing population, and this book deliberately avoids discussion of the agriculture of these areas which include most of the world's surface and population. Much has been written about the food production potential of the countries of the developing world and also on their indigenous systems of agriculture.[2]

No attempt has been made here to make a systematic study of the world's agricultural types: rather, there is a concentration on the availability of data for the study of agriculture wherever it is practised and on techniques of analysis which can be used to examine the known facts in attempting to further our understanding of agriculture. In this respect this is very much a methodological work, and many of the techniques have been borrowed from other disciplines and applied to agricultural situations. A comprehensive understanding of the available techniques of data analysis is as important as a systematic knowledge of the agricultural systems of the world; in fact the latter can hardly exist without the former. Together they can provide the basis for future progress in agricultural geography, which will not be made by further empirical studies but by discovering the right questions to ask about agriculture and by having the right information and methods of analysis with which to answer them.

CHAPTER TWO

Agricultural Location Theory

DURING the early nineteenth-century controversy concerning the current high price of grain two contrasting views were expressed. The first of these was that the price was high because of the high rent charged by landlords which had to be recouped by the farmers by charging high prices. The second view, expressed by David Ricardo[1] and others, was precisely the opposite: prices were high because of a European food shortage caused by the Napoleonic wars. Because of the high prices there was keen competition among farmers to obtain the best land for the production of such a profitable crop. This competition inflated the asking price for suitable land. The supply of land being more or less fixed, changes in the demand for it were reflected in changes in the rental charged. This analysis assumes only one product; in this example there is no possible substitute for corn and, as demand has to be met, rents must rise.

Ricardo extended this idea to the more general case where the rent for agricultural land was the result, not of the shortage of agricultural produce, but of a lack of alternative employment. Labour in this case has to compete for a limited amount of land, with few or no opportunities for employment in other sectors of the economy. Therefore rent for agricultural land becomes a function of rural population density. If there is no other employment readily available the price of land will rise according to the number of people who wish to work it.

Clark[2] has established a relationship between rural population density as represented by the male agricultural labour force and annual rent per hectare in a number of different environments ranging from eighteenth-century England to Italy in 1960. Although the situation in eighteenth-century England may have had a Ricardian element in it, it seems very doubtful whether rent is 'caused by' rural population density in modern Italy. The regression lines produced by Clark show that the two variables are related, but this alone cannot be used to support Ricardo's thesis: it appears far more likely that a third variable is causing variation in both population density and in rent for agricultural land. This third variable can be conceived as some combination of climate, soil fertility and topography. The better the land for agricultural use, the more people there are wishing to use it, and therefore demand will lead to an increase in the rent charged. This relationship will have little or nothing to do with alternative employment possibilities for the population.

From his consideration of the value of agricultural land Ricardo developed the concept of economic rent which has remained central to ideas on agricultural location, particularly those of a deterministic character. Economic rent, which need not have any relationship to actual rent, was regarded by Ricardo as the return which can be realised from a plot of land over and above that which can be realised from a plot of the same size at the margin of production. By way of illustration we can hypothesise that the return to a farmer from a hectare of wheat on land which is barely suitable for the growth of this particular crop will be £x. A return of £x + y can be gained from the growth of wheat on a hectare of highly suitable land. £y is a measure of the return per hectare over and above that which is possible at the margin of production. £y is, therefore, the economic rent. Notice that the margin of production is here determined in terms of the suitability of the land to grow the particular crop with no spatial considerations involved.

On two important points Ricardo's explanation of agricultural distributions is limited: first he considered that the margin of production is determined by factors of physical suitability only; secondly, agricultural production was considered to be homogeneous. Ricardo was primarily concerned with the production of a single, ideal crop and paid no attention to the different environmental requirements of different crops. With these limiting assumptions Ricardo's concept is clearly a normative model, deriving a pattern of agriculture from certain limiting assumptions. The suitability of the land is fixed and a pattern of agriculture results.

A second name of the greatest significance in the history of agricultural location models is that of von Thünen.[3] Although he had completed the first draft of a treatise on agricultural location before coming in contact with Ricardo's work[4] it is convenient to consider von Thünen's ideas as following from and developing those of Ricardo. Ricardo had produced a highly deterministic economic model establishing what would happen in given circumstances. Von Thünen was a practising agriculturalist with a prosperous estate near Rostock and, although he states the intention of his work to be the establishment of laws governing the price of agricultural produce, his work contains an important behavioural element as he was trying to establish why farmers behaved in the particular way that they did in his locality.

Ricardo used only two variables to account for the variations in the rent for agricultural land: the suitability of the soil and the rural population density. Transport costs, the location of markets, the demand for and varieties of agricultural produce were not considered. To a greater or lesser extent these are all included within von Thünen's agricultural writings. Although developed independently, we can regard von Thünen's thesis as substituting distance for soil fertility in Ricardo's model while at the same time allowing for a greater diversity of agricultural produce: thus transport costs are the cause and rent the consequence of differentiation in land use. Clark[5] reaches the conclusion that von Thünen's idea was relevant only to a horse transport economy, but Chisholm[6] warns against too glib a dismissal of the concepts of von Thünen who regarded his work as a method of approach to a difficult subject rather than a model to which all farming systems must approximate. The method of approach was to establish the agricultural produce needed within the urban market and the controlling factors of its production, which need not necessarily be transport costs, and to show the effects of these controls on economic rent and the pattern of differentiated agricultural production. Von Thünen's concept of economic rent is essentially the same as that of Ricardo. The crops grown in a particular place will be those which yield the highest economic rent, and the level of economic rent will be controlled by the distance to the market and the transport costs involved in getting these crops to the market. Therefore, economic rent is the income that accrues to a farmer growing a particular crop over and above that which he could expect by growing that crop at the margin of production. This margin is defined by the extra transport costs which, at increasing distance from the market, eventually add so much to the price of the produce grown that the market price is less than the production costs plus transport costs.

To illustrate these basic assumptions von Thünen proposed a series of simplifying assumptions which are typical of normative models. He assumed a uniform plain with no variations in physical factors or population density and thereby discounted Ricardo's two reasons for variations in economic rent. The evenly distributed rural population were farmers all with the same technical competence and all equally well informed as to the state of the market. This market consisted of a single city, central to the whole country and equally accessible to all the farmers, all of whom had the same method of transport—horse and cart—available to them for the movement of goods to market. The cultivated area was surrounded by unsettled wilderness with the same physical characteristics as the settled country and this unsettled area provided the room for expansion of the population and the growth of food needed for this expansion. The basic concept of economic rent in these circumstances can best be understood by considering a small settlement requiring only a small quantity of agricultural produce which, in this example, we shall take to be only grain. Suppose the city had a population of 1,000 people for whom food could be grown within a radius of four kilometres from the city. With a fixed price of £0·25 per kilogramme and a yield of 1,000 kilogrammes per hectare, the farmers next to the city would have a return of £250 per hectare. If transport costs are £62·5 per 1,000 kilogrammes per kilometre with a simple linear relationship to distance, at four kilometres from the city production would give a nil return as total financial yield would be absorbed by transport costs. This would therefore be the margin of production (Figure 1).

Farmers located at the margin of production would be advised to bid up to £250 per hectare for land next to the city. The *bid-rent* for this land is then £250 per hectare. When explaining

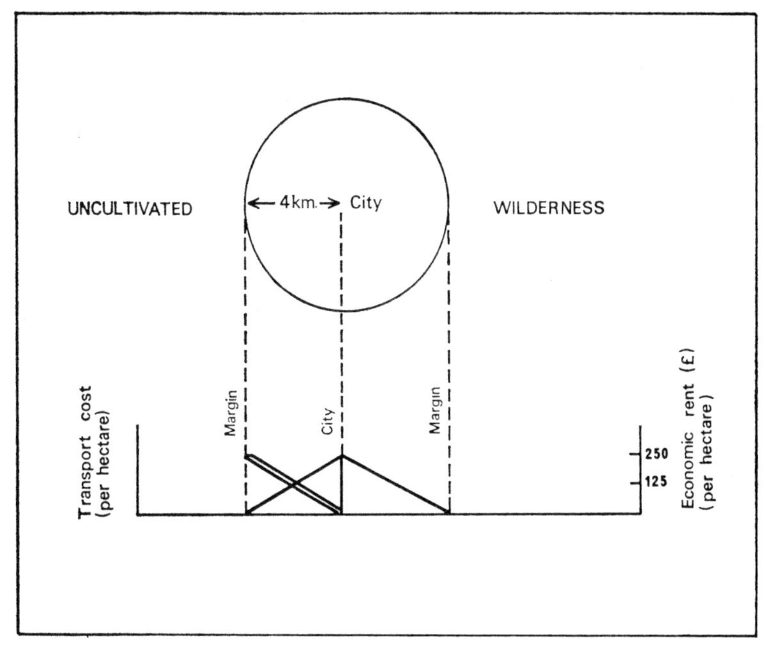

Fig 1. Relationship between economic rent and transport costs in a simple one-crop agricultural landscape with one central city.

urban land-use patterns use has been made of the concept of the bid-rent curve[7] which is very similar to economic rent as it reflects the rent over and above that paid at the margins which can be offered for a piece of land. Hoover[8] calls this the ceiling rent. The economic rent curve is the exact replica of the transport cost curve in this example as there are no other variables affecting returns to the land. If transport costs are reduced then economic rent will be increased throughout the area and returns at the margins will be higher. The margins of production will not move out because there is no increase in demand from the central city. In this circumstance all the farmers will be better off except those nearest the city with no transport costs. The economic rent will decline less steeply with distance. If transport costs rise the margins will move nearer the city and either grain will have to be grown more intensively to produce enough for the market from the smaller area or the price of grain will have to rise to allow the original area to be used. It is important to note that income of the farmer is not the same as economic rent. Farmers' income is not included in this simple model but is regarded as constant, like the fertility of the soil. All farmers are imagined to have the same needs and the same standard of living and, in this example, are all tenant farmers.

Consider what happens if the city increases its population and therefore its food requirements. If the city doubles in population, assuming no change in the method of production, twice the area of agricultural land will have to be utilised to feed this population. In our example the limit of production was at four kilometres and the area of productive land was $\pi r^2 = 50{\cdot}26$ km². The limit of production will have to extend to 5·656 kilometres in order to double the enclosed area. As the newly cultivated area is farther from the central city, greater transport costs will be incurred and the price of grain will rise. As a result the economic rent for all the land enclosed within the original cultivated area will rise. In this simple situation, with only one crop, there will be a gradual fall-off in economic rent from the central city to the margin, where it will normally be zero.

With competing crops it is clear that the decline in economic rent with distance will not be the same in all cases. This gradient depends on transport costs *per hectare* of the agricultural produce. Thus those crops which yield high bulk per unit area will incur high transport costs. This will become clear if we consider two products, timber and wheat, included by von Thünen in his original model and about which there is often confusion. In von Thünen's isolated state, where the only method of transport was horse and cart, clearly transport costs for bulky material like timber will be very high. Moreover, timber yields a high bulk per unit area so transport costs per hectare for timber are very much higher than for wheat. Therefore, despite the lack of intensity of production, timber must have a higher economic rent than wheat and will be grown near the central city, provided always that there is a sufficient demand for timber in the central city. In von Thünen's time timber was a vital resource for building and fuel so had to be grown in the isolated state: its economic rent meant that it was produced near the central city. In Figure 2 the economic rent for timber is highest between A and B while that for wheat is highest between B and C: if we rotate this graph through 360 degrees the areas of land devoted to timber and to

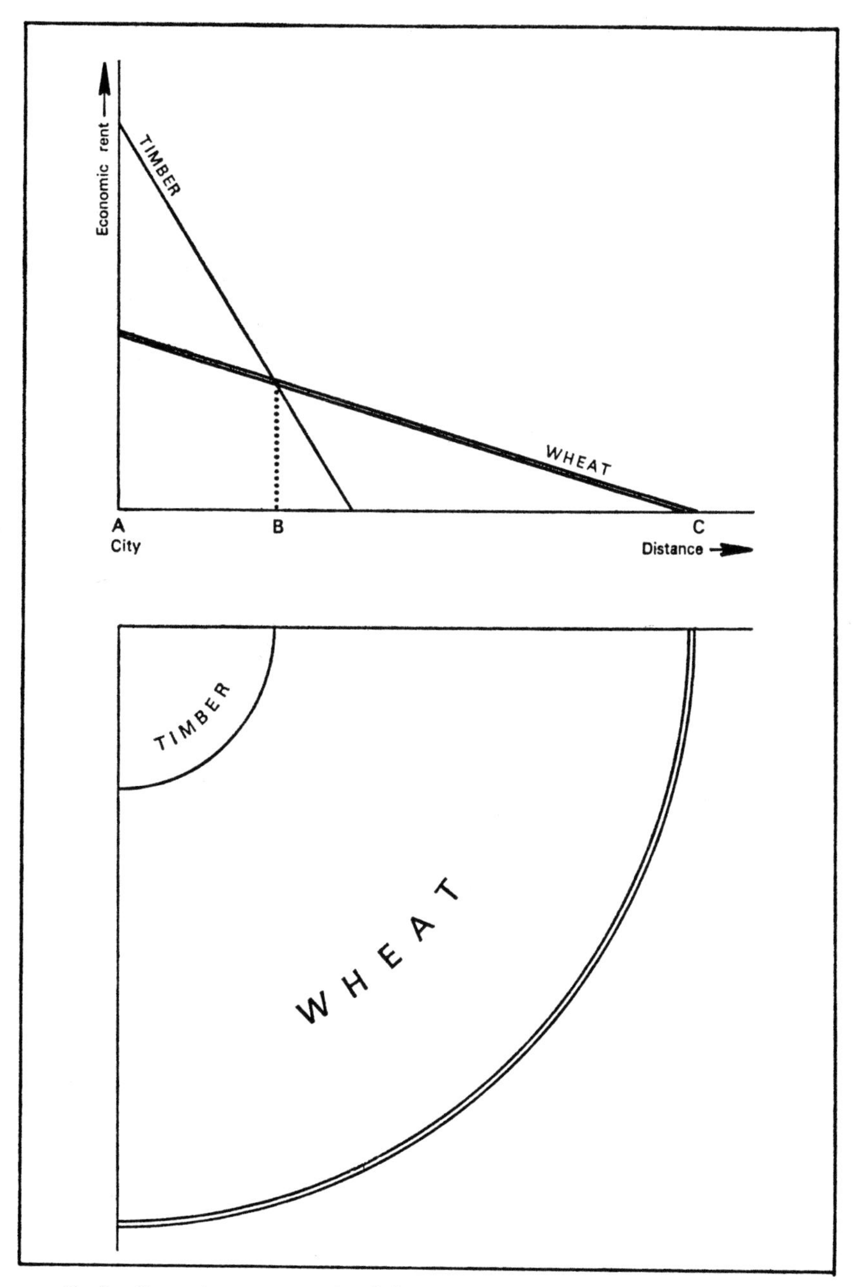

Fig 2. Economic rent curves for timber and wheat. At distance AB from the city wheat becomes more profitable than timber. Rotating these curves through 90 degrees produces the quadrant of the economic landscape shown below.

wheat appear as concentric circles. Under these conditions a different crop is substituted as transport costs rise so that bulk yield per acre falls with distance.

Alternatively, different methods of production may be substituted for transport costs so that, as these increase, the costs of various inputs are decreased. By the reduction of inputs, yields and therefore transport costs per hectare are reduced. In this way the economic rent for extensive production, although starting at a lower level, falls much less steeply than for intensive production of the same crop. To illustrate this von Thünen used three separate arable systems, growing the same crops but under different systems of intensity, which occurred in three concentric zones with the intensive arable having the highest economic rent but with a very steep rate of decline as transport costs increased. In the middle is the arable with the long ley and in the outer zone the three-field arable, the least intensive of all and therefore the system which has the highest economic rent at a considerable distance from the city.

Bringing together these two systems of substitution of crops and substitution of degrees of intensity of production for transport costs we arrive at the full pattern of land use proposed by von Thünen for his isolated state (Figure 3). Although this is a model in the sense that it can be used as an ideal to which real situations can be compared, it is more behavioural than deterministic in its construction as it rests on observations of farming behaviour and is formed to explain these.

The von Thünen model as so far expounded rests solely on transfer advantages while that of Ricardo had rested on production advantages. As with most approaches to models, von Thünen first derived a simple structure and later relaxed some of the simplifying assumptions, adding complexity and also bringing the model nearer to reality. There are three areas in which these assumptions can be relaxed: the single method of transport; the uniform plain; and the single central market. If one could cheapen and, in the case of perishables, speed up the transport of agricultural goods to market, the bid rents would decline less steeply with distance. As a result the widths of the various concentric zones would be expanded and the outer margin of production would be pushed outwards into the uncultivated area, provided there was sufficient demand for the increased product. If this transport advantage were available over only a limited area of the plain the

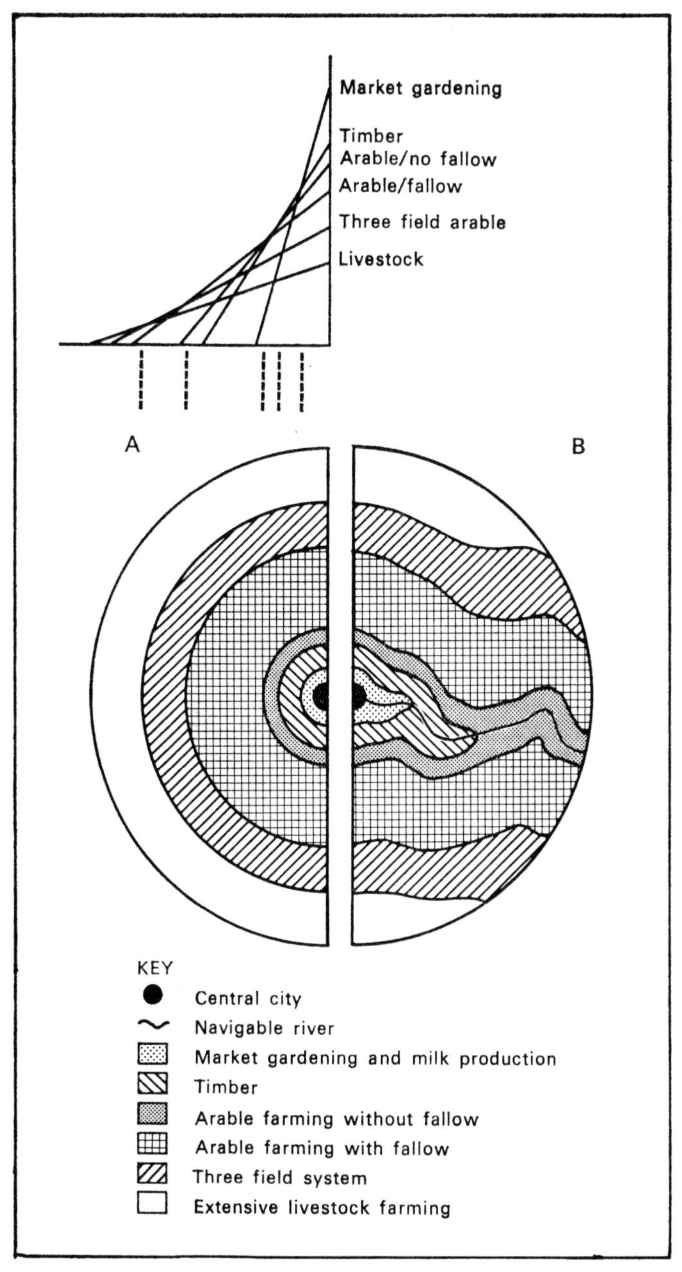

Fig 3. The full von Thünen landscape. Economic rent curves for the six agricultural systems are shown in diagram A. The effect of a navigable river on the concentric pattern of land use is illustrated in diagram B.

regularity of the concentric circles would be destroyed. Von Thünen postulated the existence of a navigable river flowing across his agricultural plain and, as he regarded the possibility of a major city not being on a river as extremely unrealistic,[9] he suggested that this river should flow through the city. With river transport costs at one-tenth of those overland the land-use pattern changes as in Figure 3. A limited example of this type of phenomenon could be found in the American Prairies after the building of the transcontinental railways. The area in which wheat was grown was limited to the area within which wheat could be carted to the stations, but the railway itself led to the extension of wheat growing for many hundreds of miles into the prairies.[10]

If the physical uniformity of the plain is relaxed so that soil fertility is non-uniformly distributed, areas of good soils will have high yields and therefore economic rent will be increased and the crop zones will be pushed outwards. Conversely, where soil fertility is low yields are reduced, as is economic rent, and the crop zones contract. The position can be made more complex if soil fertility is not the only factor but rather the suitability of the soil to grow a particular crop is varied. If transport costs are the dominant costs in production then these variations in soil fertility and suitability are likely to cause only local scale deviations from the model. As transport costs decline in importance the size and the frequency of these disturbances are increased.[11]

A further aspect of uniformity is the distribution of the farming population and their abilities, aspirations and standard of living. Von Thünen suggested that farmers living some distance from the central market, far removed from urban influence, would have less sophisticated wants and might be prepared to accept a lower standard of living. If this should be the case their margin of production could be pushed farther outwards from the city since, for the same economic return from the land, more money would be available for the payment of economic rent with less being spent on the farmer and his family. There is considerable evidence of this process in areas of marginal farming today.

The final modification which von Thünen introduced into his basic model was the inclusion of secondary market centres. As the central city grew in population and the agricultural area expanded outwards, so the growth of subsidiary settlements would be encouraged which would act as market centres in their own

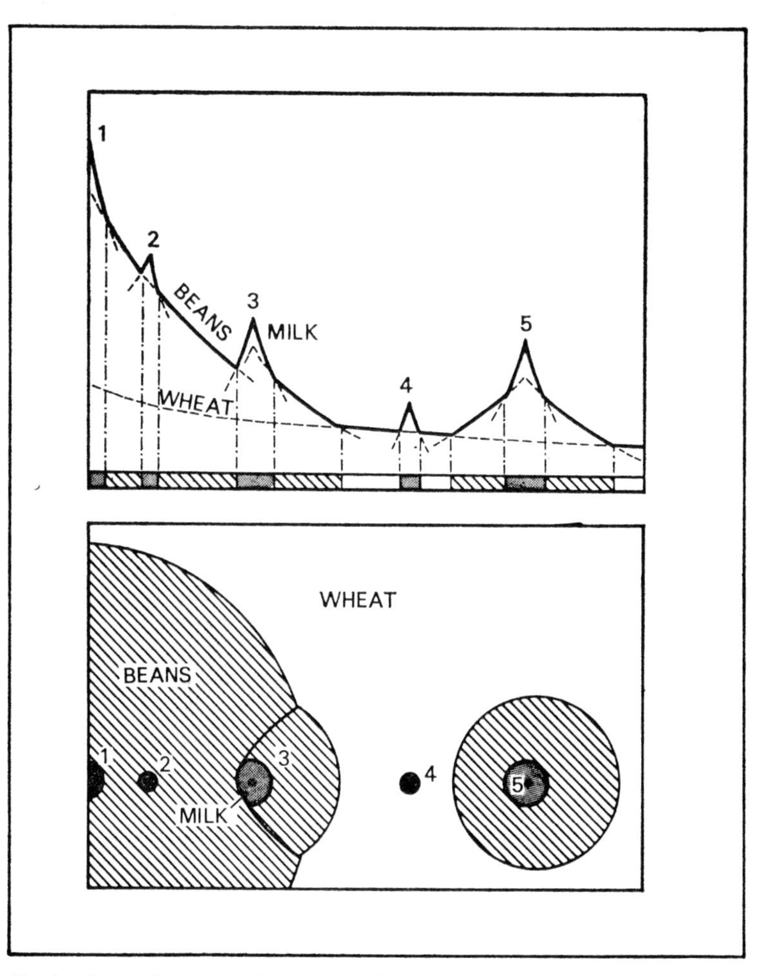

Fig 4. Economic rent gradients around five towns. A market for milk is assumed to exist at all centres and it therefore achieves the highest economic rent close to all five. A market for wheat exists only in centre 1 with a rent gradient falling slowly away from that centre. Beans are marketed in centres 1, 3 and 5. (After Hoover, E. M., *The location of economic activity*, 1948.)

right and would develop their own systems of concentric rings of agricultural land use. The interaction of the original and subsidiary centres will lead to considerable disturbance to the original pattern of land uses. Hoover[12] shows the result of rent gradients around five centres (Figure 4). For simplicity only three types of land use are considered: first, milk production, common to all centres with the highest bid rent near the centres and a steep rate of decline away from the centres; secondly, beans for which markets exist at centres one, three and five; and, thirdly, wheat with a market only at centre one.

VON THÜNEN'S MODEL APPLIED

Von Thünen's theory was based on certain propositions, many of which were the result of his own empirical observations. It is the propositions concerning transport which have changed most fundamentally since von Thünen wrote and these changes in transport do most to make his thinking appear anachronistic. The use of modern transport and refrigeration permits the movement of perishables over considerable distances and has done most to destroy the symmetry of land-use systems around central markets. Perishables can now be enjoyed in most urban markets regardless of season; the sources of supply change according to the seasonal growth pattern of the particular crop.[13] Despite this type of difficulty the method of approach to the problem remains valid. During the empirical investigation of a problem the most important variables affecting it are identified and their effects are considered in isolation from all other influences. Having first derived this 'ideal' model, other variables can be introduced into the analysis until the model situation approaches more closely that of the real world. At this stage a considerable understanding of the various controls on the problem will have been reached and the interactions of many controlling factors examined.

Despite radical changes in the assumptions underlying von Thünen's model it is still possible to see direct applications of it to real world situations today. This implies that there are certain situations where transport is still the overriding factor in the allocation of land use. This can be illustrated at three scales:

1. at the intra-farm scale in areas where no sign of von Thünen's concentric rings may be evident on a regional scale;

2. at the local level where the agricultural economy of the area has not developed to the extent of making use of sophisticated transport and marketing systems;
3. at the regional level where access to major world markets is still a factor in locating agricultural production.

Although von Thünen arranged his estate in a series of ideal sectors radiating from the centre with no likeness to his larger scale model,[14] it is possible to identify concentric patterns of land use around the centre of operations on the single farm.[15] The farm buildings usually form the centre of operations of any farm and the use to which each piece of land on the farm is put depends partly on its distance from the centre of operations. Land uses which require frequent trips by the farmer, his machinery or his livestock between the land and the farm buildings will not occupy the most remote parts of the farm. Dairy herds, for example, will tend to be grazed near the centre of operations because of the frequent contact between the land and the farm buildings, while crop raising, involving far less frequent interactions, will be found in the outer areas. An interesting variation of this case can be found where a new motorway has cut off one part of a farm from the major part, leaving one section in increased isolation from the centre of operations. The considerable expense of access roads and bridges means that not all farmers with land on both sides of the road can have their own access but must share the facilities. This might necessitate quite a considerable journey in either direction to the nearest crossing point. Often the more isolated piece of land is sold to a farm on the other side of the road or even exchanged for a piece on the first farmer's side. If such sale or exchange does not take place the isolated piece of land may undergo considerable changes in land use after the motorway is constructed, its new isolation leading to crop raising or beef cattle rather than any more intensive activity.

Several examples are given by Chisholm[16] of how village communities distribute their agricultural labour according to the distance of the various plots of land from the village. The intensity of land use changes inversely with distance, even in circumstances where the quality of the land improves at a greater distance. An example is quoted of the situation in a Sicilian village where, although the village is on top of a hill surrounded by infertile soil, the land immediately around the village is intensively tilled, while in the more fertile valleys some distance from the village, the land is either in extensive arable cultivation or is waste.

At a regional scale strong similarity with von Thünen's model can be seen. Jonasson[17] argued that north-west Europe can be regarded as one urban centre which acts as a market for much of the worldwide production of agricultural produce, especially for an inner agricultural zone of horticultural and perishable crops. Valkenburg and Held[18] showed that the average yield of eight crops in Europe follows a concentric pattern declining regularly away from this central market area: many of the reasons for this can be found in the physical character of the outer areas of Europe, as for example in western Ireland, Spain and Greece, but some at least of this decline in yields must be related to the distance from the major European market. In addition the choice of the eight crops was designed to counter the changing climatic and other physical conditions found throughout Europe so that a crop such as oats, suited to the wetter, less sunny environment of the Atlantic fringes was included with wheat, more suitable to the drier eastern parts of Europe.

Peet[19] has extended this approach to include a dynamic element. Under conditions of rising demand in the central city and reducing transport costs, there will be a rapid outward spread of farming into the uncultivated wilderness. Using data on imports into Britain in the nineteenth century Peet analyses the worldwide zones of agricultural production, especially the expansion of commercial agriculture into continental interiors of the new world in response to increased demand and rising prices in what can be taken as a von Thünen world city in Western Europe and eastern north America. By the 1870s food consumption in Britain had reached such a level that the cultivated boundary around the market had extended overseas as there was little room within Britain itself for further agricultural expansion. This outward movement was consolidated by innovations in methods of transport which later accelerated the outward expansion. The von Thünen model provides the link between the industrial revolution and population growth, which triggered off increased demand and the expansion of cultivation first in north America and later in the countries of the southern hemisphere.

Unlike most deterministic models, therefore, von Thünen's

method of study, based as it is on what happens to land use as demand grows in a central market, has a dynamic element and can be applied to situations of rapid and worldwide change with considerable success. Even though the specific assumptions von Thünen made can be questioned or altered, none the less the model can be shown to have considerable value in organising methods of approach to the study of large- or small-scale agricultural distributions, and it has formed the basis of many such studies ever since its inception.

The ideas of von Thünen have been expanded by many to a more general rent theory.[20] The typically normative approach which has to be adopted for this type of general theory is illustrated by Garrison and Marble: [21]

> For every spatial location there is some jointly optimum intensity of land use, type of land use, and group of markets, the selection of which by agricultural entrepreneurs leads to spatially ordered patterns of land use.

This has much in common with industrial location models of the least cost or maximum profit type. The difference between agricultural and industrial location theory is that the former must consider the production of a number of products while the latter is concerned primarily with one only.[22] Further, the cost of transport of raw materials is excluded in the case of agriculture, with a few minor exceptions one of which is the movement of manure from the central city for a short distance to the horticultural belt in von Thünen's model, while it must form an essential part of the theory of the location of industry.

Finally, Harvey[23] points out two areas in which the von Thünen model is in need of further development. First, as it is a partial equilibrium model, it is not necessarily easy to build into it changes over time. The assumption, for example, that changes in transport methods will lead simply to direct and obvious changes in the land-use pattern is not necessarily the case in reality. Second, there is no consideration of possible economies of scale. For example, it may be postulated that scale economies will lead to reduced production cost around the central market in comparison with the same production around a small market. If these economies of scale are sufficient to offset the greater transport cost of moving produce from the large centre to the small, then the production system of the large centre will eliminate that of the smaller centre.

Von Thünen's solution to the problem of establishing economic rent, with a demand for a variety of agricultural produce, is a maximising solution subject to a number of constraints, the most fundamental of which is that total costs of production and transport per hectare of agricultural production must not exceed the returns from that hectare. We can express the problem in the form of a linear programme.[24] If there are three zones of production around the central city and a demand for three different crops, the linear programme becomes:

$$\text{Maximise } Z = \sum_{i=1}^{3} \sum_{j=1}^{3} S_{ij}\, x_{ij}$$

where S_{ij} is the net return per hectare of the jth crop in the ith zone and x_{ij} is the area devoted to the jth crop in the ith zone. But this maximisation is subject to the following constraints:

$$1. \quad \sum_{j=1}^{3} x_{ij} \leqslant Li$$

where L_i is the total available land in the ith zone. The amount of land devoted to all crops must not exceed the total land in the ith zone. This constraint is repeated for all three zones.

$$2. \quad \sum_{j=1}^{3} C_{ij}\, x_{ij} \leqslant K_i$$

where C_{ij} is the working capital required to cultivate one hectare of the jth crop in the ith zone and x_{ij} is the area devoted to the jth crop in the ith zone: K_i is the total capital available in the ith zone. This constraint is also repeated for all three zones.

In order to maximise the first expression it may not be necessary to use all the capital or all the land available; thus the constraints, or limiting conditions, are referred to as inequalities. Here then is a classic linear programming situation with one variable to maximise subject to six inequalities.[25]

There are several other examples of how linear programming can be used to allocate resource production between different regions, and the flexibility of the technique permits the inclusion of many more restraints than von Thünen used in his basic allocation model. Heady and Egbert[26] use linear programming techniques to allocate the production of field crops (wheat, corn, oats, barley, grain sorghum, cotton and soybeans) within 122 producing regions of the United States. The criteria used for the allocation is that the lowest total supply cost should result when meeting the level of production needed to supply national needs. Supply costs were to be minimised subject to a series of constraints concerning land availability in each region for each crop. These acreage constraints were found by taking the actual acreages for the seven crops as planted in 1953. This was before the onset of production and acreage controls in the USA and, with high post-war prices, when acreages tended to be near or at a maximum. Seven further constraints were provided by the national demand for each of the seven crops, a demand which was to be met but not exceeded.

Thus, in linear programming terms we minimise:

$$Sc = \sum_{i=1}^{n} \sum_{j=1}^{m} x_{ij} C_{ij}$$

where Sc are the total supply costs and x_{ij} is the level of the jth activity in the ith region and C_{ij} is the cost per unit of the jth activity in the ith region (in this example $m = 7$ and $n = 122$).

This minimisation is subject to the following inequalities:

$$1. \quad \sum_{j=1}^{m} x_{ij} A_{ij} \leqslant S_i$$

where A_{ij} is the per unit land requirement of the jth activity in the ith region and S_i is the total land availability in the ith region.

$$2. \quad \sum_{j=1}^{m} \sum_{i=1}^{n} x_{ij} = D_j$$

where D_j is the national requirement for the commodity j.

A number of conclusions concerning the distribution of field crops in the United States results from this analysis. Should the United States government halt the present field crop stock-piling policy and change to an agricultural policy which is more in line with purely national requirements, certain changes might be expected to take place in the distribution pattern of field crop production. In particular there would be a shift away from the more marginal areas for the production of wheat, and the production of feed grains would also become more concentrated in favourable areas: cotton acreage would shift greatly from the south-east to the south-west of the United States. Some of the differences found between the ideal allocation of land resources to crop production and the present distribution of field crops may be more a reflection of the deficiencies in the data used, and in the inequality relationships established, than in the present distribution of crop production in the United States. In particular the omission of livestock from the model is a very serious shortcoming.

A further, more comprehensive, land-use allocation using linear programming has been provided by Howes[27] as part of a major study of water resource allocation by the United States Department of Agriculture: this paper is concerned with the use of linear programming as a technique for allocating agricultural production of various types to different sections of the Susquehanna river basin. In this case the objective function is to minimise the total cost of producing and shipping final crops (ie excluding fodder crops used within the region) and livestock commodities from each region of the basin to their respective market areas. The inequalities are similar to those we have already seen in the previous examples: the land used for a particular type of agricultural production cannot exceed the total available land within the region; a second inequality concerns the ratios in which intermediate goods such as fodder can be combined to produce a final good. A set of these constraints can be provided to allow for flexibility in the production functions of the final goods.[28] Another set of constraints is fixed so that the shipments of final goods from a region are all above a limited threshold. This is necessary in linear programming analysis as there is a tendency for the analysis to allocate very small productions of commodities to some regions without being able to take account of economies of scale. The final inequality is that all the shipments of agricultural commodities are of a non-negative nature. In this particular application of linear programming four producing regions and seven marketing regions are designated and the resulting optimal solution differed from the actual situation in a number of important ways. Because of these substantial differences the authors felt that the model itself was not adequate and that more work was needed to quantify further constraining relationships. On the other hand care has to be taken to ensure that constraints are not built in to the model to replicate the present position exactly; there is little benefit to be gained by simply being able to reproduce the actual pattern of agriculture. The technique has its primary use as a predictive tool for agricultural planning to show how future resources should be allocated to maximise some defined function. This type of analysis is not restricted to linear relationships, as with all the inequalities used as examples here, but quadratic and more complex functions can be used if they are empirically justifiable.[29] A most important field for linear programming is to use inequalities which deal with the reactions of farmers to possible changes in agricultural systems, especially where these changes require the abandonment of established habits in order to recoup increased returns. This takes us away from deterministic situations into the realm of behavioural studies and requires the relaxation of many of the limiting assumptions of purely deterministic models.

Two of the most basic limiting assumptions of all normative models of agricultural location, including that of von Thünen are: 1. that each individual operator has complete information; 2. that each operator makes rational decisions to maximise his returns in the light of this perfect knowledge. The extent to which these assumptions are not met in reality must be obvious to all. Farmers are expected to make rational decisions about land use without full information on yields or prices to be expected at time of harvest and sale. The extent to which this decision on land use can be said to be rational must be limited by the extent to which information is either available or can be guessed correctly and by the level of the ability of the farmer both to see the implications of each of a number of alternative actions and to carry out his chosen action. There is, in addition, the attitude of the farmer to his land. A good example of contrasting farmer attitudes is provided by Banzini[30] when comparing agriculture in the north and south of Italy. It is this dilemma which leads us away from the normative economic models of location to a study of decision-making at the level of the individual farmer. There are two methods of approach to the study of decision-making. The first of these is to derive a normative solution to the problem about which a decision has to be made, in the circumstances of a measurable degree of risk and uncertainty, while the second is to examine how the farmers behave and to relate this pattern of behaviour to the conditions assumed in normative economic models.

GAME THEORY

The normative solution to decision-making is provided by game theory,[31] a technique which has been applied to agricultural case studies.[32] Essentially a 'game' is set up whereby the farmer is playing his environment in some form. The environment has a number of gambits it can play, for example it can produce a drought, a wet year or an intermediate year. As a consequence of each of these gambits the yields of the farmer's crops are affected in a number of different ways. On the other side of the game the farmer also has a number of moves he can make. This might mean growing different crops, some of which will do well in dry years while some will do well in wet years. Similarly, game theory has been applied to the predicament of Jamaican fishermen.[33] The farther out to sea they lay their fishing pots, the larger the quantity and quality of the catch. The catch inshore is considerably less and, in times of success from the outer pots, the market is flooded and the price depressed to the extent that the poorer-quality fish from the inshore pots may not find a market at all. On the other hand the environment has some moves in this game; occasionally a current in the area of the outer pots damages the floats with the result that the pots are sunk and lost, the catch with them. There are, therefore, three alternative plays open to the fishermen. They can lay all their pots inshore with the minimum of physical risk but with the certainty of poor returns and the risk of no financial return at all at certain times. Secondly, they can lay all their pots well out to sea, taking the risk of total loss of catch and financial investment in the gear, in order to reap the higher financial returns from the successful catches. Thirdly, they can 'hedge their bets' and lay a proportion of their investment inshore and a proportion out to sea. The choice of the proportion to lay inshore depends on how the risk of failure is perceived by the individual fishermen.

In order to optimise the decision of either the farmer or the fisherman in the face of risk and uncertainty, the theory of games permits us to draw up a pay-off matrix showing the outcomes of each possible move by the farmer against each possible move by the environment. Such a pay-off matrix can be illustrated by Table 1. Although we have used actual numbers for the different crop yields under the different conditions of the environment, this need not be a quantitative measure; each play by the farmer could be ranked on a scale of comparative success or failure, these relative values being established as a result of long experience. This is the type of exercise likely to be carried out in the previous example, perhaps unconsciously, by the Jamaican fishermen.

TABLE 1 Pay-off matrix of crop yields under different environmental conditions

	Environmental conditions (rainfall)		
	High	*Medium*	*Low*
Crop 1	20	15	10
Crop 2	105	35	0
Crop 3	15	30	60

Given this pay-off matrix, there are a number of solutions to the problem of the selection of the 'best' crop. Crop 2 would give the highest yield if the environmental conditions were right, but this crop carries the risk of total failure with no yield at all if the environmental conditions are unfavourable. An alternative would be to select the crop which has the highest minimum value, that is the highest yield under the worst environmental conditions. In this case crop 3 would be chosen as the yield will not drop below fifteen. Gould[34] presents a graphical solution to this problem of a three-by-three pay-off matrix (Figure 5). This presentation shows that the choice for the farmer must lie between some combination of crop 2 and crop 3 with the latter more successful in dry years and the former more successful in wet years. In this example crop 1 gives a consistently lower yield than either of the others and in this way a three-by-three pay-off matrix can always be reduced to a two-by-three matrix. It is not necessarily, as with the Jamaican fishermen, a straight choice between one of two alternatives, but rather the selection of the best combination of each of the alternatives. Modifying Gould's method it is possible to calculate this optimum combination. We calculate the variation in yield of crop 3 under the various environmental conditions and the variation in crop 2 and calculate each as a percentage of the total variation. The proportion of the total yield variation which comes from crop 2 equals the recommended proportion

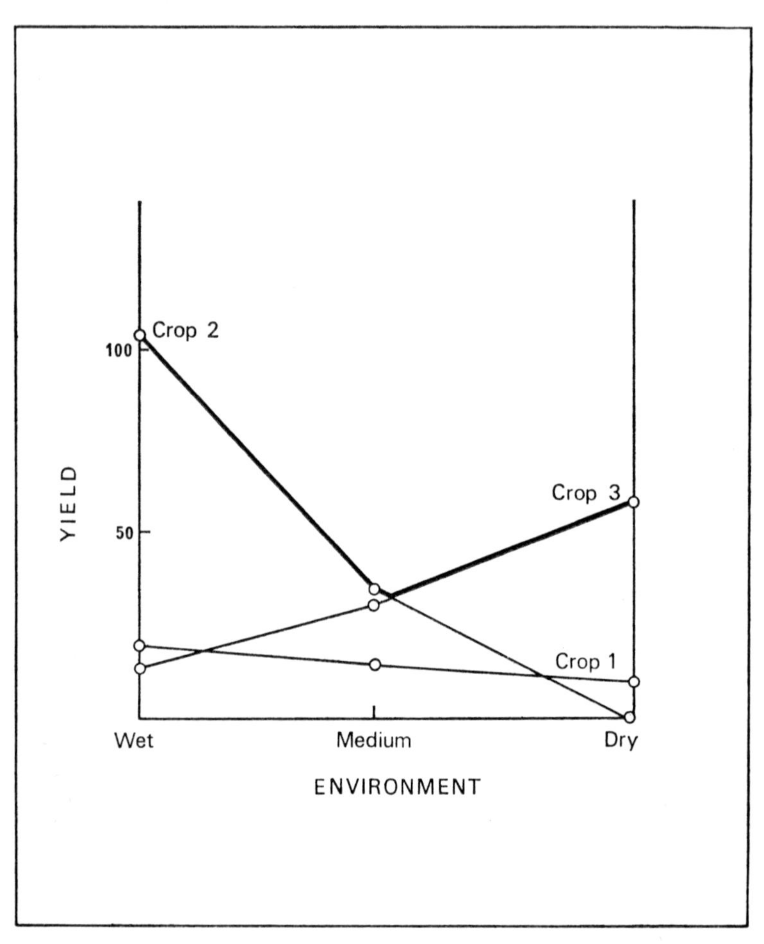

Fig 5. Representation of a three-crop, three-environment game situation. Crop 2 has a high yield under wet conditions and a low yield under dry. Crop 3 is the inverse of this and crop 1 shows a low but fairly constant yield through all environmental states. The highest yield for any given environmental state is indicated by the heavy line. In this situation the farmer should choose some combination of crops 2 and 3.

of the total effort which should be devoted to crop 3. For example, with equal numbers of wet, medium and dry years:

	Yields (x)			*Average*		*%*	*% crop*
	Wet	*Medium*	*Dry*	*yield* ($\bar{x}$)	$\Sigma(x-\bar{x})^2$	*variation*	*growth*
Crop 2	105	35	0	46·67	5,716·67	84·5	15·5
Crop 3	15	30	60	35·0	1,050·0	15·5	84·5
				Total variation:	6,766·67		

Therefore 84·5 per cent of the total variation in crop yield comes from crop 2. Crop 2 is, therefore, an unreliable crop and should not be grown in great quantity as it carries too much risk of total failure. On the other hand, the very high yields obtainable in some years mean that it should be grown whenever possible. 15·5 per cent of the total yield variation comes from crop 3, a much safer crop, but relatively unexciting in terms of yield. If we invert the percentages we have the proportions which ideally the farmers should grow of each of the two crop choices open to them. A similar exercise could be carried out for the fishermen of Jamaica.

This solution assumes that the choice between the two alternative strategies should be on the basis of the best possible worst position, the so called maximum-minimum solution.[35] In reality there are other players in the game besides the farmer and the weather, and alternative solutions to the pay-off matrix. 'Because of the time lags between investment and pay-off, the price-taking nature of agriculture, stochastic weather variables and other factors, the uncertainty faced by decision makers in agriculture is greater than and different from that confronting managers of most other sectors of economy.'[36] Taking into account all the uncertainties, there are five methods of solving the pay-off matrix.[37]

The first of these, called Wald's criterion, we have already considered. The farmer examines the worst possible outcome of each of his alternative strategies and selects the least risky. The solution is particularly appropriate in circumstances of near-subsistence agriculture where the farmer has little latitude for risk-taking but must ensure a certain minimum production in order to avoid starvation. The second solution, called the Laplace criterion, is based on the assumption that all states of nature are equally likely, which in our example means that the wet years come as often as the dry years: the farmer then makes his selection on the highest average or expected yield under each condition. The third solution, Hurwicz's optimism-pessimism criterion, is more complicated. The farmer selects a probability between one and zero for the highest pay-off for any strategy. In our example the best outcome of strategy (crop) 2 was 105. A probability of this occurring is assigned, say 0·7: 1-r is the probability (0·3) or pessimism index, which is assigned to the worst outcome of the strategy. For each strategy an index Ci is obtained by multiplying the best outcome by its probability and adding this to the worst outcome multiplied by its probability. The strategy with the highest Ci is chosen by the farmer. In the fourth method of solving the pay-off matrix, the Savage regret criterion, the farmer attempts to minimise the differences between the yield that he obtains and the yield that he could have achieved had he known beforehand the outcome of the play, that is the outcome of the gambit of the environment. In order to do this a regret matrix is created showing the difference between each outcome and the best outcome possible in that particular state of nature. Using our previous example a regret matrix can be created:

		Environment		
	Wet	*Medium*	*Dry*	*Total regret*
Crop 1	105 − 20 = 85	35 − 15 = 20	60 − 10 = 50	155
Crop 2	105 − 105 = 0	35 − 35 = 0	60 − 0 = 60	60
Crop 3	105 − 15 = 90	35 − 30 = 5	60 − 60 = 0	95

From this matrix we select the crop which gives the minimum regret, in this case crop 2. The fifth alternative is a compromise between Wald's criterion and the regret criterion; it can be called the benefit criterion. For each state of nature we select the worst play by the farmer. All other plays for this state of nature would increase the benefit. For this we create a benefit matrix.

		Environment		
	Wet	*Medium*	*Dry*	
	worst play 15	*worst play 15*	*worst play 0*	*Total benefit*
Crop 1	20 − 15 = 5	15 − 15 = 0	10 − 0 = 10	15
Crop 2	105 − 15 = 90	35 − 15 = 20	0 − 0 = 0	110
Crop 3	15 − 15 = 0	30 − 15 = 15	60 − 0 = 60	75

If the maximum-minimum strategy is applied, no one crop has the advantage but in terms of total likely benefit the farmer would select crop 2.

Wald's criterion suggests complete ignorance on the part of the farmer, who is planning for the worst but has no idea how likely this is to occur. This policy is best for the subsistence farmer or the one who will not take even the slightest risk. The Laplace model suggests that the farmer is aware of the extreme states of nature but has no idea how often they strike: he therefore adopts the law-of-averages solution on the basis that he will achieve average yields over the long term. On the other hand he is able to risk short-term reduced yields to achieve this average position. Agrawal and Heady[38] suggest that this solution is appropriate for an experienced farmer who moves into a new area, the environment of which is strange to him. The Hurwicz model is the reverse of the Laplace model in that it is not concerned with the average position but takes into account only the probability of occurrence of the extreme positions. As it rests on farmer judgement of these probabilities it is likely to give erratic results. The regret solution is less pessimistic than Wald's criterion and will sometimes lead to a choice which will give higher yields. The benefit model is a compromise between the Wald method and the regret method.

Game theory gives us a rather cold-blooded, deterministic solution to the problem of explaining reactions to a state of risk. It is wrong to consider the environment as a player in this sense. The farmer can make more or less rational choices based on his experience while the environment varies in what may, over the long term, appear to be either a systematic or a random manner. It is not, we assume, making the same judgements as the farmer. The environment is not, then, a true opponent, as only the farmer wins or loses in this game. This is not a strict application of game theory and the pay-off matrix cannot be interpreted in terms of any pay-off to nature in the same way as it is interpreted as pay-off to the farmer.

The concept of uncertainty, introduced through game theory, means that, regardless of the actual motivation of farmers, an optimising goal cannot be reached. If the farmer is uncertain regarding the outcome of the weather, or some other influential variable, he will almost certainly have to be satisfied with some level of attainment which is less than the optimum. Given the circumstances of the individual farmer, it is unlikely that he is in a position to even know what the optimum solution to his range of choices would be, even if he could control such uncertain factors as the weather. As a result it would appear more satisfactory for us to examine the ways in which the farmer achieves what to him is a satisfactory level of returns on his farm. This satisficer solution will be found at widely differing levels of productivity depending on a number of factors which can be divided into two groups: the level of knowledge of the farmer and the level of uncertainty or risk in the production (Figure 6).

Uncertainty itself stems from a degree of instability in the factors affecting farming while the consequences of this uncer-

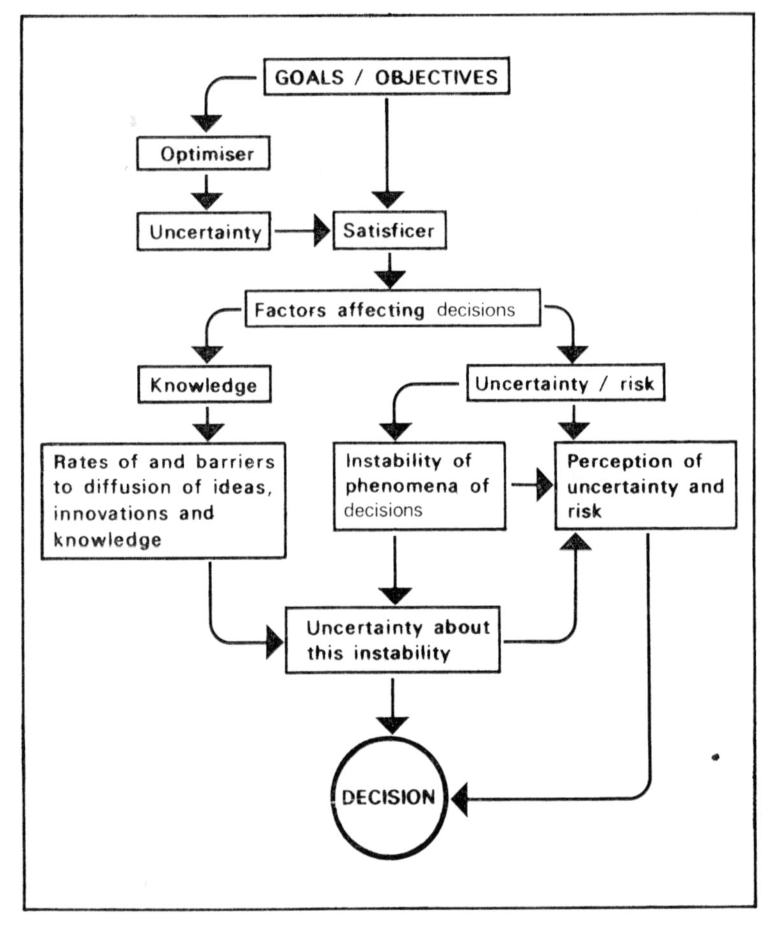

Fig 6. The effects of uncertainty and risk on decision-making.

tainty may be mitigated or aggravated by the farmer's perception of the magnitude of the risk and by his attitude to the taking of risks. Instability in the factors about which the farmer has to make decisions may operate in a number of ways. Climate is the first of these and that which has already been considered with reference to game theory. In a way it is the most severe type of uncertainty, as it is that over which the farmer has no direct control: although he can mitigate some of the effects by diversifying the crops grown, he can have little or no control over the effects of extreme events such as floods, tornadoes, or total drought. It is possible for society to offset the effects of climatic uncertainty through government subsidies and compensation for full or partial crop loss but this is rarely as satisfactory as either a successful harvest of that particular crop or a successful harvest of an alternative crop.

A second factor which may affect the outcomes of farmers' decisions, although it does not always affect the actual making of those decisions, is disease of crops or livestock. Despite the catastrophic effect of a foot-and-mouth outbreak on a dairy farmer, the possibility or risk of this disease is likely to have little or no bearing on the decision of the farmer to adopt dairying as a major or as a subsidiary activity. The extent to which such a risk is accounted by a farmer will depend on many factors, especially on his past experience. Uncertainty introduced by regular outbreaks of disease will have a greater or lesser effect depending on the assessment of the likelihood of occurrence, the cost, availability and knowledge of techniques to mitigate the disease, and the extent to which the market prices reflect the degree of risk involved. The level of uncertainty which is introduced by disease and which can be tolerated is a highly variable factor.

The market itself provides a further cause of uncertainty for any producer, and the farmer, with substantial time lags between preparation and marketing, will feel this uncertainty more than most. Demand can fluctuate for three reasons: first, over- or under-production by farmers will lead to falling or rising prices; secondly, demand can vary independently of price fluctuations. Apparently random changes in taste can occur which can have dramatic effects on prices, although this is unlikely in the short term in agriculture. Thirdly, government policy, or lack of it, on import restrictions and control can add substantially to the problems created by over-production at home and lead to or help prevent further depression in the market. There are various ways in which uncertainty introduced through demand fluctuations can be mitigated. A government or purchasing company can fix a minimum market price for a commodity or can control production to prevent gluts and therefore depressed prices. On the whole, the uncertainties brought about as a result of market structure are more predictable, less catastrophic, and can be mitigated more easily than those resulting from the weather.

Other areas of potential instability in the factors of production are evident on the farm itself. Perhaps the most obvious concerns the health of the farmer and the effect this might have on his ability to continue to operate his farm in the way he planned at the start of the season. Similarly, there is the stability of his labour force. The departure of the best stockman and the lack of a suitable replacement at a time of labour shortage on the land might have serious effects on the profitability of the major farm enterprise. Other family changes may affect a farmer's attitude to innovations and risks. Such are some of the contributing factors making a farmer either a conservative or an innovator. But other factors such as education also play a considerable part.

All these factors increase the uncertainty of the circumstances in which the farmer makes decisions. They introduce instability which prevents the attainment of an optimising goal, even if such a thing could be identified. The uneven pattern of these levels of uncertainty is one reason for variations in farming practices over an area such as the United Kingdom. In addition farmers are not all equally well equipped to deal with this problem of uncertainty since they are not equally well informed of the levels of risk, of the methods of mitigating the risk, nor are the facilities for such mitigation equally distributed. This applies especially perhaps to the capital necessary to implement diversification, disease prevention measures, mechanisation and so forth. The spread of information through the farming community will be considered separately but it is clear that not only the uncertainty, but also the ability to deal with this uncertainty, must provide some of the explanation for variations in farming patterns.

Even more important than the actual risk and uncertainties of farming is the farmer's perception of this uncertainty. This involves not only the ability of farmers to recognise the risk for what it is but also their reaction to risk and the multitude of factors which influence their willingness to take risks with capital and with income stability. The study of perception in geography,[39] which has achieved importance recently, is of considerable relevance to decision-making in agriculture. Clearly a decision which involves a fixed amount of capital will be regarded very differently by a hobby farmer in East Anglia[40] and by a farmer in the Welsh highlands who has farmed with little change in technique for the last thirty years or more. The perception of risk will be very different in each case. The recognition and impact of risk or uncertainty will vary with many factors, one of which will be the age of the farmer. There is a widely held view that it is the young farmers who are the innovators or risk-takers but this need not necessarily be so. The young farmer, who has often little capital to risk and who may have a young family to support, is not often in a position to be a risk-taker. We have also to set a lack of experience in practical farming against a recent and up-to-date training in farming methods. On balance the young farmer would seem more amenable to change and would perhaps not regard particular change as risks at all, in contrast to more traditional farmers. Two other factors affecting attitude to risk-taking have already been considered, family size and structure and working capital. A gambler with unlimited capital and no family to support will normally have different attitudes to stakes than one who requires a regular income to feed and clothe a family. Thus a small farmer in Western Ireland, with little or no capital and a large family acquired rather late in life[41] establishes a farming method which is ultra-conservative in the face of an unpredictable environment. Instead of relying on beef or sheep raising such farmers derive a substantial proportion of their income from dairy cattle, the principal advantage of which is not the magnitude of the return from the investment, but the regularity and frequency of what return there is. The greatest difficulty is experienced by anyone who attempts to initiate change in such an agricultural environment. Many have noted how experience is the all-important factor both for and against change: whereas existing residents may rely on their experience to support them in their conservatism, newcomers from another area, bringing with them their own traditions and experience, will normally be the greatest protagonists for change. The migration of farmers from the central valley of Scotland to Suffolk and north Essex in the late nineteenth century, bringing their dairy cattle and sheep into a predominantly cereal-producing area, provides a useful example.[42] Other, wider-scale, examples are to be found in the colonisation of Malaysia and the importation of rubber trees from South America, and in numerous other cases throughout the world of new farming introductions through colonisation.

THE DIFFUSION OF INNOVATIONS

The suggestion that it is the newcomers to an area who lead to the adoption of new ideas takes us to the wider consideration of the processes by which ideas and innovations spread through a farming population and what encouragements and barriers there are to this spread. If one of the reasons for the complex pattern of farming is the geographical variability of unstable factors of production, another is certainly the uneven rate of the diffusion of ideas, solutions to particular problems, information of government assistance and many other things through the farming population.

As with perception, there have been very extensive studies of diffusion by geographers in recent years.[43] This series of investigations had its origins in the work of Professor Hägerstrand at the University of Lund.[44] In order to simulate the spatial diffusion of ideas and innovations Hägerstrand established a number of rules, and all factors not included within the scope of these rules were allowed to act in a random manner in their influence on diffusion. While individual behaviour cannot be considered random, the minor factors which influence decision-making in a large number of people can collectively be expected to operate in a random manner, some cancelling others out so that no overall direction in their influence can be determined. This random element permits us not only to allow for the operation of all the many variables affecting man's actions which we have not been able to include in the rules, but also allows us to consider some of man's actions in a non-rational, non-economic way.

The first, and the most important, rule to be established is the mechanism by which information is passed from one individual to another. In the case of information among farmers Hägerstrand assumed that this would be by face-to-face contact and that the likelihood and frequency of such contact is determined by the distance apart of the two individuals concerned. Some distance-decay function has then to be established to replicate the way in which contact decays or declines with distance. This has become one of the fundamental rules for simulation of diffusion and, although the decay function may be changed, the basis on which it is derived has remained the same.

Working on the adoption of innovations in farming in the Asby district of Sweden, Hägerstrand calculated this distance-decay function of personal contact using details of known migration distances for the local population. The number of people migrating from Asby falls off rapidly with increasing distance from Asby. Thus at a short distance away a large number of people have moved from the town, at increasing distance this number declines according to the empirically derived function:

$$Y = \frac{6{\cdot}26}{D^{1{\cdot}585}}$$

where Y is the number of migrating households per square kilometre and D is the distance of the mid-point of this square kilometre from Asby.

Assuming that the possibility of social contact is the same as the possibility of migration, this equation can be used to plot the Mean Information Field[45] from Asby showing the degree of likelihood of a person at an outside location receiving face-to-face contact with someone at the centre of the field. By dividing the area around Asby into square cells, each at known distances from Asby, a value for Y can be calculated for each cell. By dividing each cell value by the summation of the values for all cells the probability of contact with each cell can be estimated. Having thus established the probability of contact between one centre and all others, this lattice is used as a floating grid which can be centred over any cell which has a teller ready to pass his information on to a potential adopter. With the help of random numbers it is possible, using this lattice of probability values, to select the grid cell to which the information is passed. For the second step in the simulation two cells have tellers and the floating grid is moved so as to centre over each of these in turn.

Using the spread of adoption of a Swedish farm subsidy scheme, Hägerstrand was able to compare the actual diffusion from 1930 to 1932 with a series of simulated diffusions. Such a comparison, cell by cell over the grid, illustrates how the actual and simulated patterns of diffusion differ and suggests ways in which the set of rules established for the diffusion simulation can be improved. The most obvious possible source of error is the distance-decay function. A new exponent for distance may have to be used if the nature of the mean information field is different from that which was predicted. Also, as the distance-decay function is rotated through 360° to give the mean information field, the decay of contact with distance is assumed to be the same in all directions from the centre of the floating grid. Geographical barriers to diffusion can be simulated by making contact between neighbouring cells more difficult by allocating lower probability values to them. In this way the simulated pattern of diffusion can be brought closer to the real one; hence a greater understanding has been reached concerning the ways in which ideas are spread and the effects that the uneven spread of ideas has on the geographical patterns of farming (or on other activities).

In a study of the adoption of pump irrigation on the high plains of Colorado, Bowden[46] notes that new wells seem to decrease in number away from the location of earlier wells. The nearer a farmer is to a neighbour with a pump installation, the greater the chances that he will hear about it, observe the operating well and then adopt the same irrigation methods himself. The mean information field was established, not by migration data as in Hägerstrand's study, but on a two-tier basis. The local level of contact was established using the distance travelled and attendance at a local free barbecue. On the Colorado plains, with few barriers to communication, a larger-scale flow of information was recognised. In conjunction with the barbecue data, telephone calls over ten miles were used to satisfy both the local and the regional scales of communication. A floating azimuthal grid was established and placed in turn over each existing well and, because there is a natural limit to the number of wells which can be successfully operated in a local area, a township was withdrawn from the simulation when it had more than sixteen wells. The simulation illustrated consistent similarity to the actual pattern of well locations.

THE DIFFUSION CURVE

It is clear from an examination of the pattern of adoption of pump irrigation in Colorado that there is a characteristic end state to the diffusion process which is achieved when the innovation has spread from the original centre or centres to cover the whole area of potential adopters, thereby saturating the market for the innovation. During the later stages of a diffusion we can expect a gradual slowing down to an eventual halt of the adoption process. During the early stages of the diffusion of an innovation there will be a certain resistance to the new technique resulting in a rather slow start to its dissemination. As the innovation is adopted by more and more people, so there are more and more to pass on the idea and there is a rapidly accelerating rate of adoption which begins to slow down only with the onset of saturation of potential adopters. If adopters are plotted as a cumulative percentage of all potential adopters with time a characteristic curve results. The precise shape of the curve depends on the nature of the innovation and the degree of resistance expressed by potential adopters. This curve can be plotted as a normal distribution of the number of adopters with time. Using standard deviation units of this distribution, Rogers[47] suggested a fivefold division of adopters on the basis of the length of time during which potential adopters have to be exposed to the innovation before they accept it (Figure 7). Anderson,[48] in a study of agri-

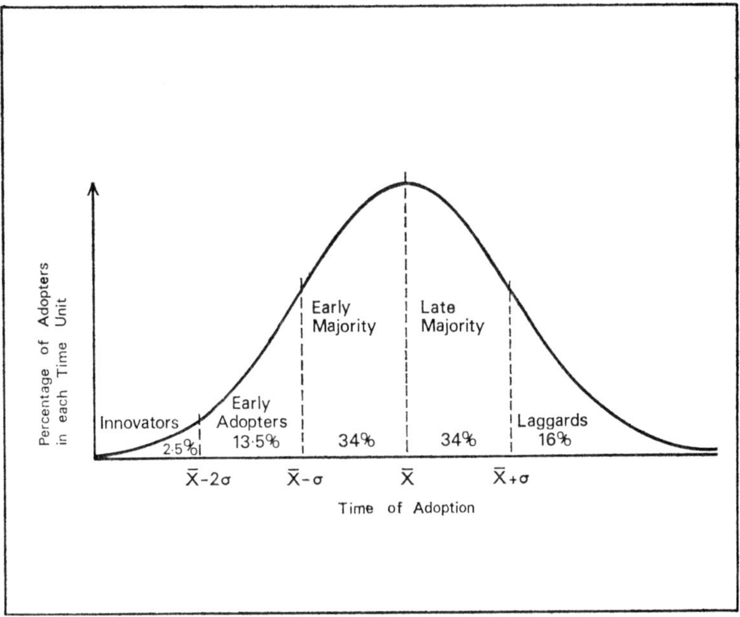

Fig 7. Distribution of adopters of an innovation with time. (After Rogers, E. M., *The diffusion of innovations*, 1962.)

cultural innovation in Denmark, finds that labour-saving techniques are adopted almost instantaneously whereas more expensive investments, to which there would be naturally a greater resistance, follow the diffusion curve more closely. In addition there is clear evidence that the rate of diffusion has increased in recent years and that the time period needed to reach 50 per cent adoption of an innovation has decreased considerably.

In an interesting study of the adoption of hybrid corn in the United States Griliches[49] shows that not only were there differences in the rates of acceptance of the hybrid corn among farmers in different parts of the United States but that once the corn had been introduced into an area the rates of adoption varied also. Although the particular strain of corn used in this example had been developed in Connecticut, commercial development started first in the heart of the corn belt where the market for

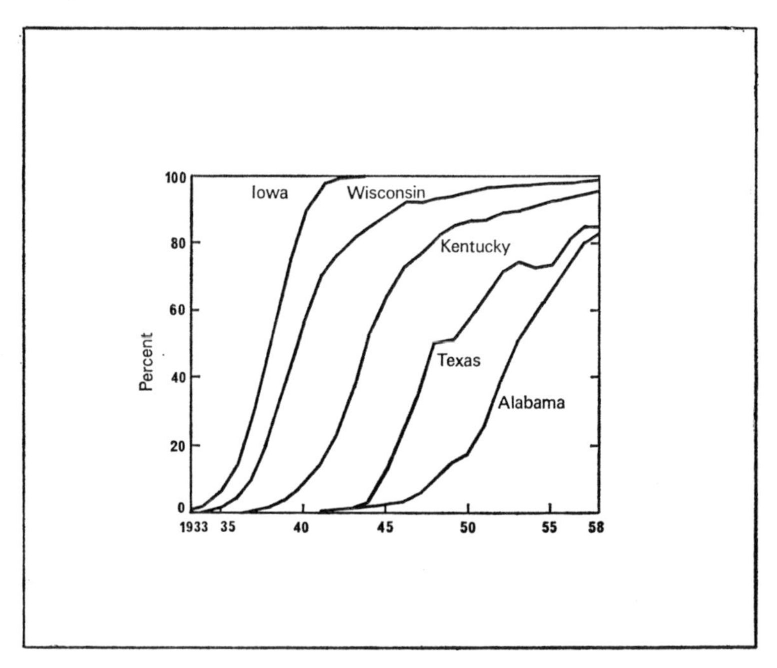

Fig 8. The diffusion with time of the adoption of hybrid corn seed in five American states. The corn was adopted first and fastest in the main corn-growing states of Iowa and Wisconsin. (After Griliches, Z., 'Hybrid corn and the economics of innovation', *Science*, 132 (1960), 275–80.)

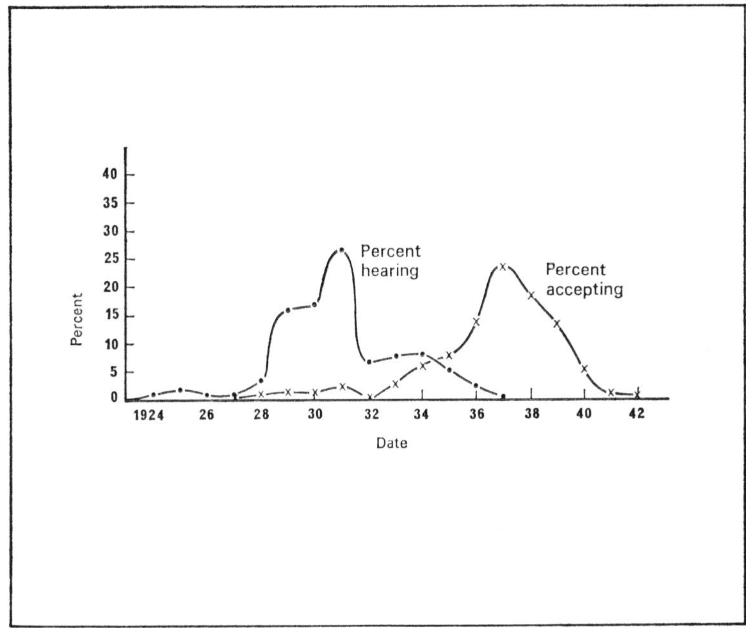

Fig 9. The difference in the diffusion with time of hearing about and adopting the innovation of hybrid corn seed in Iowa. (After Ryan, B., and Gross, N. C., 'The diffusion of hybrid seed corn in two Iowa communities', *Rural Sociology*, 8 (1943), 15–24.)

the seed was largest. From there it spread outwards to the margins of corn production: farmers in Iowa showed a much earlier acceptance of the new seed than those in Texas or Alabama. Iowa showed the fastest spread and the south-eastern states and the Mississippi delta states the slowest (Figure 8). This was not the result of a limited supply of the seed but rather the difference in the magnitude of the demand for it, a difference in demand which is, to some extent, a reflection of differences in the magnitude of profit to be gained from adopting the innovation. The rate of acceptance showed a striking relationship to the corn yields in the different states; the higher the yield of the hybrid corn the better the return to the innovation and the faster the adoption rate. In an earlier, more local, study of the same hybrid corn seed Ryan and Gross[50] were able to compare the shapes of the diffusion curve and the time scales of the diffusion of knowledge of hybrid corn and the diffusion of its adoption (Figure 9).

The modal values of the two distributions are separated by six years and although the seed was first heard about in 1924, it was not until 1937 that actual adoption reached a peak. The pattern of the adoption of the corn does not follow a normal distribution; there are less adopters in the earlier years and more than expected at, and just after, the modal year. In examining the major influences listed by the adopting farmers as affecting their decision to adopt, the authors were able to show differences between the innovators and early adopters, and the late majority and the laggards. The earlier adopters were most influenced by seed salesmen while their more conservative neighbours awaited the local success of the seed before trying it themselves and listed their neighbours as the most important influence on their decision. Typical diffusion curves can be found for the diffusion of new agricultural machinery (Figure 10); similar results have been reported in many other types of study.[51]

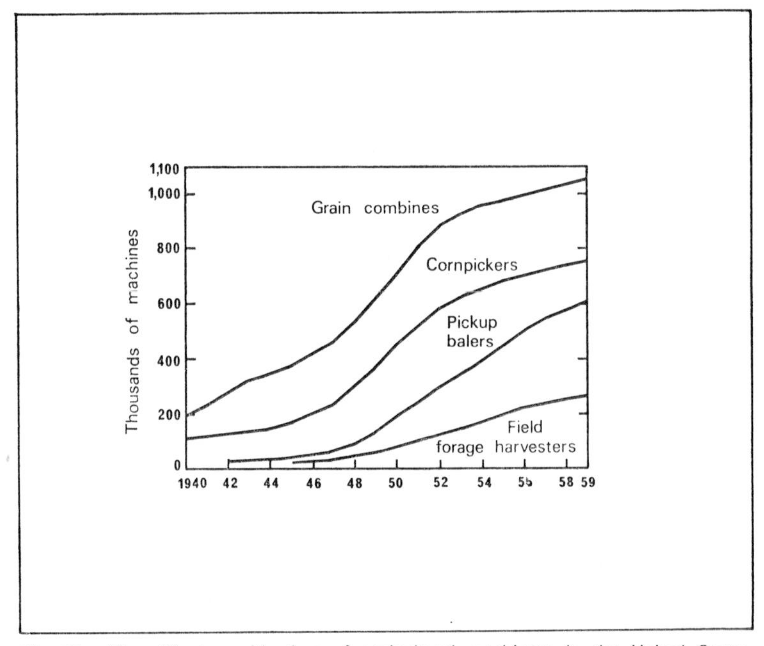

Fig 10. The diffusion with time of agricultural machinery in the United States. (After Griliches, Z., 'Hybrid corn and the economics of innovation', *Science*, 132 (1960), 275–80.)

Pred[52] has shown that simulation of diffusion may do no more than mirror the current real-world position. In any such stochastic process the operations will have to be replicated several times over in order to arrive at a stable 'average' result. This average result may bear a strong resemblance to the real-world situation but it is difficult to assess the amount we have learnt about this situation purely as a result of the simulation exercise. This question takes us right back to the first assumption when setting up the simulation procedure, the nature of the mean information field. Most observations of the information field are substitutes for the actual process which is to be simulated which may have little justification beyond an apparent empirical 'rightness'. There is a danger in all such experiments, where the results are known, that the inputs to the experiment can be so modified as to give the predicted result so that little or no further knowledge is gained about the process under study. It is also quite possible to so establish a simulation exercise as to achieve the correct results for totally the wrong reasons and to have little idea of the extent of this illusion.

A further doubt over the value of the simulation of diffusion of ideas is in its discontinuous representation of a continuous process. First-generation adopters tell second-generation adopters who in their turn tell third-generation adopters. This process can be continued for as many steps as are needed to replicate the real-world situation or until the problem becomes too great for the computer to handle. The continuous process of diffusion must be separated into a number of stages which have no parallel in reality. Finally there is the important question of the assessment of the success of the simulation process in a particular case. The difference between the predicted number of adopters in each cell and the actual number can be compared by using the Chi square distribution measuring the significance of the differences between the expected number of adopters—the actual pattern—and the observed number—the simulated pattern. This method becomes a crude generalisation when a large number of grid cells are examined and consistent local deviations between the actual and the simulated patterns may have little effect on the overall significance test. Such comparisons between the actual situation and the various simulations are usually made by eye.

Monte Carlo simulation procedures need not be confined to attempts to simulate diffusion patterns. In any circumstances where there are a number of alternative choices, each with a given probability of selection or of success, possible future positions can be simulated with the help of the selection of random numbers.[53] Such an application is suggested by Donaldson and Webster[54] as an alternative to linear programming for farm planning. In seeking to maximise profit by the selection of the most profitable combination of possible farming activities, linear programming has a number of drawbacks which a simulation technique can attempt to overcome. Subjective factors and preferred solutions can be weighed. Discontinuities associated with indivisible resources such as labour and machinery, and individual units of production, such as livestock, can be avoided. Upper and lower limits to the size of each activity can be specified to prevent the proliferation of a multitude of small units which may be included in the solution proposed by linear programming, taking little account of fixed costs.

Using random numbers a sample of all possible plans is drawn; each is tested for feasibility by making sure that the resources of the farm are sufficient for the plan and that the level of each of its activities is within the limits specified. Plans can be weighed: for example, if the farmer is known to have a preference for dairy cattle all plans with a considerable proportion of dairy activities can be given a greater chance of being sampled. If it is found feasible the plan is expanded so that it uses all the available resources of the farm and from these plans a sub-sample of the twenty most profitable is selected. Each of these solutions will be at or near the maximum level of profitability. This type of search and maximisation procedure requires considerable resources of computer time but the results may have more meaning to farmers, and therefore have more practical value, than linear programming methods.

MARKOV CHAINS

The simulation of the diffusion of innovations is concerned with reproducing spatial distribution patterns with increments of time. This process assumes only a one-way change. As an innovation may be adopted or ignored the farming pattern may either change in response to an outside stimulus or remain the same. Once an innovation has been adopted, it is possible that it may later be rejected in favour of a further development or in favour of the position before the innovation was introduced. No longer do we

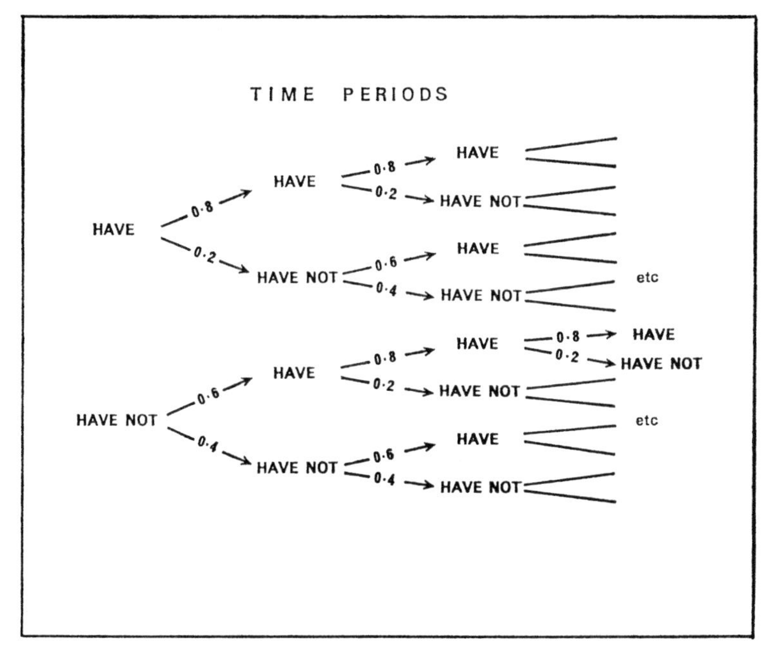

Fig 11. Representation of transition probabilities of a change in state among farmers having or not having adopted an innovation.

have only the probability of the adoption of an idea; we have also the probability of a negative change in the system. There may also be other changes possible if we introduce other innovations at the same time. In order to investigate the outcomes of such a large number of alternatives, some more probable than others, we use a technique called Markov chain analysis.[55]

In the place of the mean information field probability values we assign a probability value to a change from an initial, or starting, state to all other possible states. In the diffusion examples we had two possible states: that of adopting the innovation and that of not adopting it. Knowing the probability of an adoption taking place we can draw up a *transition matrix* (P_{ij}) showing the probabilities of a change in state from a pre-existing state (i) to a future state (j).

		Succeeding state (j) *Having*	*Not having*	*Total*
		innovation		
Existing state (i)	having	0·8	0·2	1·0
	not having	0·6	0·4	1·0

After one time period there is a 20 per cent chance that past adopters of the idea will abandon it.

The probabilities of changes in state can be represented as a branching tree for more than one time period (Figure 11). Therefore by multiplying the probabilities of change we have:

$$(0{\cdot}8 \times 0{\cdot}2) + (0{\cdot}2 \times 0{\cdot}4) = 0{\cdot}24$$

the probability of a change from having the innovation to not having it after two time periods. Doing the same for all possible changes of state we have the following transition matrix for two time periods.

$$(2:1)\ \mathbf{P}_{ij}\ (T_o - T_2) = \begin{pmatrix} 0{\cdot}76 & 0{\cdot}24 \\ 0{\cdot}72 & 0{\cdot}28 \end{pmatrix}$$

A quicker solution to this problem can be achieved by powering the original transition matrix. If, for example, we wish to know the transition matrix after two time periods we square the original transition matrix:[56]

$$(2:2)\ \mathbf{P}_{ij}^2 = \begin{pmatrix} 0{\cdot}8 & 0{\cdot}2 \\ 0{\cdot}6 & 0{\cdot}4 \end{pmatrix}^2 = \begin{pmatrix} 0{\cdot}8 & 0{\cdot}2 \\ 0{\cdot}6 & 0{\cdot}4 \end{pmatrix} \cdot \begin{pmatrix} 0{\cdot}8 & 0{\cdot}2 \\ 0{\cdot}6 & 0{\cdot}4 \end{pmatrix}$$
$$= \begin{pmatrix} 0{\cdot}76 & 0{\cdot}24 \\ 0{\cdot}72 & 0{\cdot}28 \end{pmatrix}.$$

If we suppose that the initial state of the innovation is such that 30 per cent of the population have already adopted it and the remaining 70 per cent have not, we can represent this situation by a probability vector:

$$(0{\cdot}3 \quad 0{\cdot}7).$$

The state of the innovation after two time periods can then be found by multiplying this vector by the square of the transition matrix:

$$(2:3)\quad (0.3 \quad 0.7) \cdot \begin{pmatrix} 0{\cdot}76 & 0{\cdot}24 \\ 0{\cdot}72 & 0{\cdot}28 \end{pmatrix} = (0{\cdot}732 \quad 0{\cdot}268).$$

After two time periods 73 per cent of the farmers will be using the innovation and 27 per cent will not. After a number of iterations (Pn time periods) this process will reach a steady state, or equilibrium, where there will be no change in the probability vector with time, or rather, any change in one direction will be balanced by a comparable change in the reverse direction. In the above example, if we multiply the original transition matrix by its square we arrive at the transition matrix after three time periods. Multiplying the result by the probability vector gives the following probability vector after three time periods:

$$(2:4) \qquad (0{\cdot}3 \quad 0{\cdot}7) \cdot \begin{pmatrix} 0{\cdot}76 & 0{\cdot}24 \\ 0{\cdot}72 & 0{\cdot}28 \end{pmatrix} \cdot \begin{pmatrix} 0{\cdot}8 & 0{\cdot}2 \\ 0{\cdot}6 & 0{\cdot}4 \end{pmatrix} = (0{\cdot}746 \quad 0{\cdot}254).$$

Alternatively we can multiply the probability vector after two time periods (2 : 3) by the original transition matrix to achieve the same result:

$$(2:5) \qquad (0{\cdot}732 \quad 0{\cdot}268) \cdot \begin{pmatrix} 0{\cdot}8 & 0{\cdot}2 \\ 0{\cdot}6 & 0{\cdot}4 \end{pmatrix} = (0{\cdot}746 \quad 0{\cdot}254).$$

After three time periods the rows of the transition matrix are very similar and are nearly the same as the resulting probability vector. Two further iterations yield:

$$(2:6) \qquad \begin{pmatrix} 0{\cdot}8 & 0{\cdot}2 \\ 0{\cdot}6 & 0{\cdot}4 \end{pmatrix}^5 = \begin{pmatrix} 0{\cdot}750 & 0{\cdot}250 \\ 0{\cdot}750 & 0{\cdot}250 \end{pmatrix}$$

and the resulting vector equals:

$$(2:7) \qquad (0{\cdot}3 \quad 0{\cdot}7) \cdot \begin{pmatrix} 0{\cdot}750 & 0{\cdot}250 \\ 0{\cdot}750 & 0{\cdot}250 \end{pmatrix} = (0{\cdot}750 \quad 0{\cdot}250).$$

The transition matrix now has identical rows which are also identical to the resulting probability vector. No further iterations are necessary as they would continue to yield the same result. The process has reached a steady state. Seventy-five per cent of the farmers will be using the innovation and 25 per cent will not after five or more time periods. There will continue to be change in the system, as some farmers will reject the innovation and some will adopt it, but the proportions with and without will remain the same.

Unfortunately this achievement of the steady state, although mathematically neat, is unlikely in agriculture as it assumes that the original transition matrix remains the same throughout the operation of the model. It is more likely that, as farmers change from one state to another or from one type of farming to another, the future probabilities of other farmers making changes of the same type are reduced. This situation is made clearer if one of the states of the system, or types of farming, is *absorbing*. This means that once a farmer has changed to a certain type of operation he will remain in that state and will, therefore, have no probability of change to any other state with succeeding time periods. We can imagine such a situation where the capital investment for making a change to a particular type of farming is so great that the probability of a change into this state is low but, once it has been made, the large capital investment will ensure that there is no possibility of a change out of this state. If we have three states of farming A, B and C with a transition matrix P_{ij}:

		Succeeding state j			*Sum of probabilities at succeeding state*
		A	B	C	
	A	1	0	0	1
Initial state i	B	0·2	0·6	0·2	1
	C	0·1	0·5	0·4	1

this can be represented diagrammatically as:

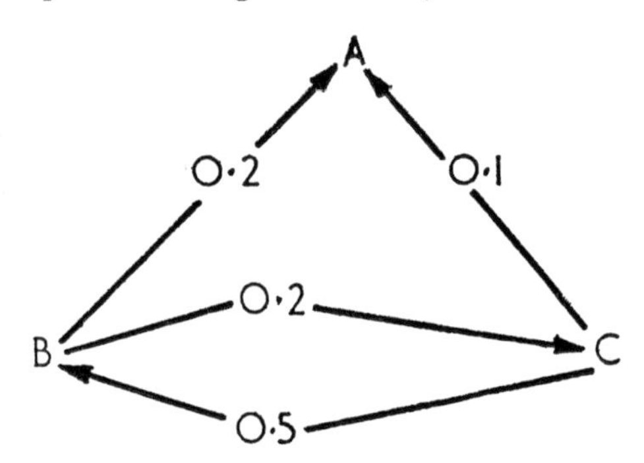

Although the probability of change to state A is low, eventually all farmers will be in the absorbing class. When this is the case the system will have reached a steady state. In the real world, part of the incentive for, and the probability of, a move from one type of farming to another will be the lack of the supply of the produce of type A and therefore its high price. As more farmers make the change to type A in response to this incentive, so the deficiency is made up, prices decline and the possibility of a move out of type A arises. With time the transition matrix needs to change.

The same situation can be illustrated if we use Markov chains to analyse the changes in the use of land over time.[57] The probability of some land changing to urban use from agriculture may be quite high in an expanding urban situation. Urban land use is likely to be almost solely an absorbing class: once the land has made this conversion it is most unlikely to revert to agriculture. If we allow such a resulting Markov chain process to continue to a steady state the inevitable conclusion is reached when all the land is in urban use. What is a more likely process is that as more and more land is converted the probability of further change in land-use is reduced; hence the steady state is so delayed as to be virtually unattainable.

As the same steady state is reached regardless of the original allocation of farmers among classes, the model is deterministic. The same input transition matrix gives the same conclusion. It is possible to make Markov chains into stochastic processes. Returning to the previous case of three types of farming with the transition matrix:

		Succeeding state j		
		A	B	C
	A	1	0	0
Initial state i	B	0·2	0·6	0·2
	C	0·1	0·5	0·4

with a probability vector of

$$(0{\cdot}2 \quad 0{\cdot}5 \quad 0{\cdot}3)$$

we can predict the allocation of farmers between the three states by drawing random numbers. If there were one hundred farmers in all, the twenty in type A would remain there throughout because A is an absorbing class. The fifty farmers of type B could be allocated between the three possible types by drawing random numbers such that:

any numbers between 0·000 and 0·1999 represent changes from B to A;

any numbers between 0·200 and 0·799 represent farmers not making a change;

any numbers between 0·800 and 0·999 represent changes from B to C.

Thirty random numbers could be drawn for farmers in type C such that:

any numbers between 0·000 and 0·099 represent changes from type C to A;
any numbers between 0·100 and 0·599 represent changes from type C to B;
any numbers between 0·600 and 0·999 represent no change.

This process can be repeated for as many time steps as are required. Unlike the matrix powering method, this stochastic process would not achieve the same result after a fixed number of iterations or time periods. In this respect the techniques become more akin to the Monte Carlo techniques used for the simulation of diffusion.

One of the criticisms of the Monte Carlo simulation technique is that it represents a continuously changing process by a number of discrete steps. The same applies to the Markov chain methods. We represent continuous change by $T_1, T_2, T_3, \ldots T_n$ steps. In many cases it would be useful to replace the transition matrix P_{ij} with a rate of change matrix Q_{ij}. Clearly it should be possible to derive the rate of change matrix from the probability matrix as T, the time interval, approaches zero. As we are dealing only with the rate of change from a state i to a new state j, we can ignore the probability of a farm remaining in the same state.

M_i or the total rate of change of a state can be derived from:

$$M_i = \frac{-\log_e P_{ii}}{\Delta t}$$

where P_{ii} is the probability of no change in state during time t. A difficulty occurs in the estimation of Δt in this equation. Krumbein,[58] when applying the Markov chain technique to sequential deposition following marine transgression, assumes that deposition rate is directly related to time and that therefore Δt could be estimated as the unit of time needed to deposit five feet of sandstone and can be considered as 1·0. Drewett[59] takes a similar view: a 10 per cent increase or decrease in urban land requires a fixed unit of time Δt. If, in our example, a 10 per cent change in type of farming requires a fixed interval of time we can set Δt equal to 1·0. Clearly this time period for 10 per cent change in farming will be affected by many variables and is unlikely to remain the same throughout long periods of time but a changing Δt could be introduced as a later complication to the method.

We can now produce a column of the summed rates of change out of each of the initial states where Δt is equal to 1·0 and P_{ii} are provided by the principal diagonal of the transition matrix P_{ij}:

Initial state	M_i (rate of change out of initial state)
State A	$\frac{-\log_e 1{\cdot}0}{1} = 0$
State B	$\frac{-\log_e 0{\cdot}6}{1} = 0{\cdot}511$
State C	$\frac{-\log_e 0{\cdot}4}{1} = 0{\cdot}916$

The total rates of change out of each state now have to be sub-divided into change into the two other states. For example, the M_i for state A will be divided into the rate of change out of A into B and out of A into C. The off-diagonals of the Q_{ij} matrix can be derived from the equation:

$$Q_{ij} = \frac{\dfrac{P_{ij}}{M_i}}{P_j}$$

where $i = j$ and P_j is the total probability of a move out of state i.

So in our example:

		Succeeding state j			
		A	B	C	M_i
	A	X	0	0	0
Initial state i	B	0·256	X	0·256	0·511
	C	0·153	0·763	X	0·916

The rate of change out of a state to another state allows us to calculate the waiting time in any state before a change takes place.

$$t_c = \frac{\log_e (1/U)}{M_i}$$

where U is a random number between 0 and 1.

The greater is M_i or the total rate of change out of a given state, the smaller will be t_c or the waiting time.

Suppose a farmer had started operations in type C. His rate of change out of this category is provided by M_i where i = C =

$$\frac{-\log_e 0{\cdot}6}{1} = 0{\cdot}511.$$

If we draw a random number (say 0·7132) the waiting time for that farmer in type C is provided by

$$\frac{\log_e (1/0{\cdot}7132)}{0{\cdot}511} = 0{\cdot}6615 \text{ of the time periods used.}$$

The state to which the farmer changes will be given by

$$\frac{Q_{ij}}{M_i}$$

$$\text{for example, from C to A} = \frac{0{\cdot}153}{0{\cdot}916} = 0{\cdot}167$$

$$\text{from C to B} = \frac{0{\cdot}763}{0{\cdot}916} = 0{\cdot}833.$$

There is therefore a 17 per cent probability that he will change to type A and an 83 per cent probability that he will change to type B after 0·6615 time intervals. Using random numbers between 0 and 1 such that 0·000 to 0·1699 denotes a change to type A and 0·1700 to 0·9999 denotes a change to type B, the state of farming can be predicted after set time intervals.

Markov chain techniques have been applied to actual situations of change in agriculture with considerable success. On the other hand they have two properties which it must be reasonable to assume hold for the agricultural data being used. The first of these is the constancy of the transition probability matrix P_{ij} throughout all time periods. As any attempt to predict from the model rests on this important assumption, the external forces acting for change in agriculture will have to remain the same throughout the prediction period. The second property of Markov chains is their ergodicity, or tendency to move towards a state of dynamic equilibrium. It is important that the system being examined can reasonably be expected to tend towards dynamic equilibrium. Colman,[60] in a study of the changing structure of the dairy industry of north-west England, shows how this industry is essentially stable. This stability is imparted by four characteristics:

1 the lack of suitable alternative agricultural activities on the predominantly heavy clay soil;

2 the fact that 34 per cent of the farms are less than fifty acres in size and need to concentrate on intensive activities like dairying;
3 the milk pricing policy which has been in operation since 1931 leads to a stability of returns to dairy farming;
4 the amount of capital investment tied up in a dairy herd and the related equipment gives a considerable inertia to the industry.

The Markov chain analysis was found to be a relatively good way of predicting changes in the size structure of the north-west dairy industry, although limitations concerning the sample size used meant that the prediction was weakest in the size groups of holdings where the number of farms in the sample was particularly small.

A more general application of Markov chain techniques is provided by Power and Harris[61] in an attempt to predict changes in the structure of farming in England and Wales using seven categories of farm type based on the information published annually by the Ministry of Agriculture. The transition probability matrix (Table 2) is subject to a degree of uncertainty depending on the length of run of previous annual records which are used to provide the transition probabilities, but the high values on the principal diagonal shows the high degree of stability of the farming types. Using this transition matrix the authors projected the number of farms in each group for 1969 and compared these projected data with the actual data collected by the Ministry of Agriculture for that year. The percentage variation between the predicted and the actual data is shown in Table 3. The same transition matrix was used to project to 1973 and the resulting number of farms in each group are shown in Table 4. The projected decline in the numbers of farms in the mixed category provides a good indication of the continued specialisation in British farming. The disadvantage of the technique in assuming, and in fact requiring, stability in all factors of change remains a serious drawback. While all prediction techniques must rest to a greater or lesser extent on the assumption of continuing past trends into the future, Markov chain methods without a stochastic element are more restricting than most.

This chapter has considered a number of ways in which it is possible to investigate agriculture. On the one hand we have

TABLE 2 Transition matrix (1964–8) for different UK types of agriculture

	All dairying	*All livestock*	*Pigs and poultry*	*All cropping*	*Horti-culture*	*Mixed*	*Holdings of under 275 smd*	*Exits*	*Tota*
All dairying	0·8999	0·0120	0·0023	0·0048	0·0007	0·0241	0·0394	0·0168	1·0000
All livestock	0·0224	0·8331	0·0022	0·0156	0·0007	0·0291	0·0399	0·0570	1·0000
Pigs and poultry	0·0217	0·0065	0·7892	0·0145	0·0076	0·0257	0·1091	0·0257	1·0000
All cropping	0·0055	0·0090	0·0061	0·8884	0·0147	0·0256	0·0298	0·0209	1·0000
Horticulture	0·0032	0·0019	0·0034	0·0214	0·8586	0·0105	0·0768	0·0242	1·0000
Mixed	0·1266	0·0676	0·0169	0·0654	0·0114	0·6462	0·0324	0·0335	1·0000
Holdings of under 275 smd*	0·0107	0·0060	0·0049	0·0039	0·0073	0·0014	0·9175	0·0483	1·0000
Entries	0·0325	0·0208	0·0181	0·0234	0·0281	0·0085	0·8686	—	1·0000
All types	0·1847	0·0743	0·0278	0·0791	0·0466	0·0451	0·5055	0·0369	1·0000

(After Power, A. P., and Harris, S. A., 'An application of markov chains to farm type structural data in England and Wales', *Journal of Agricultural Economics*, 22 (1971), 163–77)

* Standard man-days

TABLE 3 Projected and actual numbers of farms by type, 1969

	Actual total June 1969	Projection 1 totals June 1969	Percentage departure from Actual
All dairying	51,951	51,936	−0·03
All livestock	21,601	20,939	−3·06
Pigs and poultry	9,022	8,260	−8·45
All cropping	24,120	25,425	+5·41
Horticulture	12,989	13,883	+6·88
Mixed	13,361	11,461	−14·22
Small holdings under 275 smd	113,166	116,633	+3·06
All holdings	246,210	248,537	+0·95

models and techniques of analysis which search for optimum solutions to the location of different types of agriculture. Von Thünen's model and the technique of linear programming allow an optimal allocation of resources within constraints laid down by the initial assumptions concerning costs and the demand structure of the market. Given the extreme complexity of the decision-making environment of agriculture it is extremely improbable that such optimal solutions are in any sense realistic. Information and ideas tend to spread erratically through a population and the state of agriculture at one moment is partly a reflection of what it was previously and partly the result of the outcome of certain probabilities of change. The inappropriateness of optimal models and the recognition of the importance of probability in agricultural location and change emphasises the importance of stochastic methods of analysis in agriculture where uncertainty and 'noise' can be introduced through the operation of random variables. Such concepts as diffusion and such methods as Markov chain analysis use probability in this way. No matter what method is chosen in a particular case, its adequate testing will depend on the possibility of collecting and interpreting adequate data. The following chapters will, therefore, concentrate on agricultural data and their interpretation.

TABLE 4 *Projection of the movement of holdings by type of farming between 1969 and 1973*

	All dairying	*All livestock*	*Pigs and poultry*	*All cropping*	*Horti-culture*	*Mixed*	*Holdings of under 275 smd*	*Exits*	*Total*
All dairying	35,044	2,107	488	1,230	349	2,559	8,987	1,187	51,951
All livestock	1,776	10,713	238	1,147	227	1,170	5,321	1,009	21,601
Pigs and poultry	678	244	3,556	422	229	405	3,202	286	9,022
All cropping	851	795	446	15,313	1,059	1,245	3,802	609	24,120
Horticulture	293	164	157	831	7,151	296	3,717	380	12,989
Mixed	3,445	1,664	421	1,794	389	2,659	2,570	419	13,361
Holdings of under 275 smd	4,548	2,374	1,746	1,973	2,875	802	93,859	4,989	113,166
Entries	173	92	65	100	114	37	2,346	131	3,058
Projected totals, June 1973	46,808	18,153	7,117	22,810	12,393	9,173	123,804	9,010	249,268

(After Power, A. P., and Harris, S. A., op cit, Table 2)

CHAPTER THREE

Agricultural Data Sources

THE collection and publication of data on many aspects of national life is regarded as one of the important functions of central government in most countries of the world. These data are made available to other government departments and often form the basis of the planning and provision of many government services. The independent collection of information on a large scale by all government departments, even if it were feasible, would be extremely wasteful of time and resources. In addition, such centrally gathered statistics often become available for private research work to both educational institutions and commercial interests.

The best-known of these sources of data which are collected centrally is the national population census. The population census contains a great deal more information than a simple count of the population at a particular time: it contains information on family size, employment, housing, level of education and many more details of the people.

Many countries also have a census of agriculture. In some countries, for example the United Kingdom, this census is organised by the government department directly concerned with agriculture, in this case the Ministry of Agriculture, Fisheries and Food, while in others, for example the United States, it is a part of the service of the central Bureau of the Census. As with the population census, a census of agriculture starts by collecting information about small individual units, usually farms, amalgamating the information gathered into larger areal statistical units. Data are summarised and published for these larger units and for the country as a whole. The use of such statistical units, rather than those for which the data was originally gathered, leads to considerable problems in the use and interpretation of the data which will be considered later in this chapter.

A full-scale population census is held, usually every ten years, in most countries but the time interval for the agricultural census varies from country to country. In the United Kingdom and Sweden it is held annually while in the United States agricultural censuses are held quinquennially. This difference in the time intervals between successive enumerations leads to some difficulty in making comparisons between different countries.

THE UNITED KINGDOM AGRICULTURAL CENSUS

In the United Kingdom, agricultural statistics have been gathered since 1866 and, like all censuses, these statistics report the state of agriculture in the country at a single time in the year, which has now become established as midnight on 4 June. The acreage of crops, the numbers of livestock and all the other information returned is correct for this single time in the farming year. Although the original returns are destroyed, county summaries are available for every year since 1866 and parish summaries for most of these years.

The first problem with a census of agriculture is to decide what exactly constitutes a farm and therefore which of the millions of landowners or tenants should be included in the census as practising agriculturalists. As the basis of the census is agriculture rather than argriculturalists, the unit of enumeration is the agricultural holding rather than the person employed in agriculture. In the 1969 United Kingdom census the definition of a farm was altered to eliminate the quite substantial number of holdings previously listed as farms but having little or no agricultural activity.[1] Since 1892 returns have been made for holdings containing over one acre of land used for agriculture, that is, devoted to crops or grass or rough grazing: about one-sixth of these had little or no agricultural activity. Since 1969 approximately 47,000 holdings with less than ten acres of crops and grass, no full-time labour and not more than twenty-five standard man days of production (p. 133) have been excluded from the census. At the same time holdings of less than one acre are now included if they represent intensive agricultural holdings, for example, in some horticultural activities. The Ministry of Agriculture is investigating the problems involved in the identification of these holdings. The information requested on the questionnaires sent to farmers has varied considerably over the years as successively more and more information is demanded of them to suit the changing pattern of farming in the nation and also to eradicate ambiguities and errors which have come to light over the hundred or more years that the returns have been made. This naturally leads to difficulties when studying the changing agricultural pattern of an area and will be considered in more detail later.[2] The most recent changes have been to bring the Agricultural Census for England and Wales into line with the World Agricultural Census produced by the Food and Agricultural Organisation of the United Nations. The measurement of the size of the total holding will now include the land under such features as buildings, roads, etc, not simply the total acreage of crops and grass and rough grazing. Information on land tenure, dropped in England and Wales after the census of 1922 although continued in Scotland, will be reintroduced. Further, questions on farm labour will be expanded to include the farmer himself and also managerial, book-keeping and maintenance staff in order that a closer estimate of the total labour force employed in the agricultural sector may be made.[3]

The returns in recent years have followed a broad division into three sections. The first and largest section is concerned with land use within the farm. The area devoted to each crop is listed to the nearest quarter of an acre (0·618 hectares) including rough grazing, woodland and other land that is not in agricultural use. Many of the categories are subdivided to permit considerable detail; for example, information is recorded on the acreage of sixty-one horticultural crops. The second section concerns the agricultural animals on the farm, the major divisions being between cattle, pigs, sheep and poultry. In the case of cattle, distinction is made between cows and heifers in milk and calf and between those cattle forming the dairy herd or destined for the dairy herd and those reared for beef. There are additional questions concerning the number of intensively reared beef cattle, the number of Irish stores and the number of calvings between

March and May of the year of the return. Pigs and sheep are subdivided by sex and age and poultry by type, with hens and pullets divided by age. The final section lists the number, sex and age groups of the workers on the farm, indicating whether their employment is whole-time, part-time or seasonal.

The parish summaries are made available to the public for a small charge. These are now produced on pre-printed computer output paper, each parish being conveniently compressed on to a single sheet. All the information collected on the questionnaires is reproduced and the farms in the parish are subjected to some preliminary grouping analysis. A table of size groups of holdings is prepared listing thirteen size categories. In addition farms are analysed by the amount of labour required on the farm using an index based on labour requirements measured in terms of standard man days which will be discussed in Chapter 4. The farms in the parish are also classified into one of fourteen categories according to the main type of farming practised. Although the method of classification could be improved upon (see Chapter 4) it does provide some preliminary analysis of the raw statistics for the purchaser of the parish summaries. If summaries for a large number of parishes are required it is possible to purchase them on computer magnetic tape copied from the Ministry of Agriculture's own records which are held in this form. This greatly facilitates the analysis of large amounts of data.

For exercises involving the use of statistics for the whole of England and Wales the parish may be considered too small a unit. The Ministry produces summaries for Agricultural Development and Advisory Service (ADAS) districts* and for counties. The ADAS districts are amalgamations of between thirty and forty parishes for each of which there is an agricultural advisory office. There are about 350 of these in England and Wales making them convenient units for analysing the agriculture of the whole country.[4]

In addition to the annual returns published in the form of the parish summaries, each farmer makes one other return per year. One-third of the parishes of England and Wales are sampled three times a year, each sample including about 80,000 holdings. The dates of these samples are March, September and December, and

* In 1971 the ADAS replaced the National Agricultural Advisory Service (NAAS).

similar information is collected to that included in the June returns. It was suggested that after 1970 the size of these samples could be reduced to about 30,000 through a statistical stratification of the farms, with little or no reduction in the accuracy of the estimates made from these smaller samples.[5]

THE UNITED STATES CENSUS OF AGRICULTURE

The United States Census of Agriculture differs from that held in the United Kingdom in a number of important ways. The most important of these differences is that it is held quinquennially rather than annually. Less frequent enumeration is compensated for by collecting substantially more information than in Britain. The differences in the returns for the two countries are to be found in the greater scale of the United States in terms of distance and, more directly, in terms of the number of farmers. This greater scale is also reflected in the far more diverse agricultural systems found in the United States which require a greater degree of flexibility in the census procedure.

The United States census of agriculture was first held in 1850 and then decennially until 1920. From 1920 to 1950 it was held quinquennially. The next census was in 1954 and it has been taken quinquennially since, the most recent being held in 1969. One substantial contrast between the censuses of America and Britain is the inclusion in the former of data on yield of crops, usually expressed in terms of weight and also cash yield. To gather this information the date of enumeration is set so as to fall after the major harvest in each area of the United States. The actual date, therefore, varies from state to state: in 1964 the data were collected between 9 November and 23 November, with the exception of Alaska, where there is a considerably shorter growing season than in the contiguous states, for which the date was set as 5 October.

The fifty states are divided into 22,899 enumeration districts and within each the census is conducted by a combination of a postal questionnaire and personal interview. The questionnaires are delivered by post some two to three weeks before the census day to approximately eight million rural postbox holders. Not all these are farmers, so whether or not the owners of these boxes are included within the census depends on a follow-up interview. The enumerators are not sent into the field with the definition of a farm as used by the census, but rather are asked to make certain that a form is returned from all of these box holders in any one of the following circumstances:

1 the owner regarded his landholding as a farming operation;
2 there are one or more cattle on the land and/or four or more pigs and/or thirty or more chickens and/or thirty or more turkeys, ducks or other poultry;
3 there is any hay, tobacco, grain, or other field crop grown;
4 there are twenty or more fruit trees, grape vines, or planted nut trees;
5 there are any vegetables, berries, nursery or greenhouse products which were *grown for sale*.

Later computer editing of the returns eliminates those which do not comply with a more rigorous definition of a farm as set out in Table 5. Prior to 1950, enumerators were given a definition of a farm and interpreted it differently and, therefore, some at least of the changes in the number of farms prior to 1950 can be attributed to differing standards of enumeration.

TABLE 5 US Census of Agriculture: definition of a farm (1970)

Ten or more acres of land plus one of the following:

2 or more acres of crop failure
5 or more acres of crops or improved pasture
10 or more acres of other pasture
5 or more acres of summer fallow
50 or more chickens over four months old
5 or more hogs or pigs
5 or more cattle or calves
2 or more milk cows
0·2 acres or more of tobacco
100 pounds or more of tobacco produced
0·5 acres or more of vegetables or berries for sale
0·5 acres or more in orchards, vines, or nut trees
3 or more acres of harvested hay
2 or more acres of corn (provided no pigs or hogs)

If less than 10 acres of land then a farm if:

5 or more acres of crop failure
100 or more chickens over four months
10 or more hogs or pigs

10 or more cattle or calves
4 or more milk cows
0·3 acres of tobacco
500 pounds or more of tobacco produced
2 or more acres of vegetables, berries etc
2 or more acres of fruit trees
5 or more acres of corn (if no hogs or pigs)

The content of the questionnaire is varied depending on the region of the United States in which it is being used. Because of the geographical diversity of the country an excessively large questionnaire would be needed if all the crops grown anywhere in the United States were to be included on the questionnaire sent to every farmer, regardless of his location. For this reason every state has its own questionnaire, and there are two such forms for the state of Texas. However, despite the variations in the individual crops and livestock included, the questionnaires are all divided into the same fourteen sections, although the questions in the last five of these are completed only for a sample of the farms. Sections one and two list details of the acreage of the land operated for agriculture by the respondent, including details about whether this land is owned, rented, managed, or leased to others. Computer checking then ensures that the same parcel of land is not recorded by more than one respondent.

Section three concerns crops that were, or will be, harvested in the year of the census. Different crops are mentioned in different parts of the United States; as an illustration, in the state of Nebraska details are recorded of eight field crops: crops likely to be grown in only small quantities are recorded to the nearest tenth of an acre and most questions refer to the acreage harvested, the harvested weight and the weight sold. In certain cases, where the actual weight sold is not easy to measure, as in the case of vegetables, the farmer is asked to record the value in dollars of the sale of all such crops on his land. In other cases the value of production can be computed, without asking the respondent for financial details, by multiplying yield by current prices.

Section four, entitled 'land use and irrigation this year', includes information on the acreage devoted to the production of crops, including hay; acreage devoted to pasture; acreage used only for soil improvement grasses, cover crops, and legumes that were not harvested or grazed; acreage of land on which the crop failed and was not therefore harvested; acreage of woodland grazed and ungrazed; acreage in houselots, barns, roads, etc. This information, which is not yet contained in the British returns, allows the total acreage returned to be checked against the known total acreage for the country or the state. This provides an indication of the extent of double- or under-recording. The questions on irrigation concern the total acreage irrigated and, in the case of the Nebraska census, the acreage of pasture and the acreage of corn and soybeans which were irrigated and the yield expressed as harvested weight from these irrigated crops.

Section five concerns the production and sale of various forest products. Section six provides demographic and economic information about the farmer and his family. Questions are included on race, age, place of residence, date of starting operations on the farm, the number of days on which the operator works off the farm and the amount of money earned providing hunting, fishing, picnicking, camping, lodging, and other recreational services on the farm.

Section seven concerns livestock, and section eight is devoted to dairy cattle. Questions concern the number of cows and heifers milked on the day before the census, the number in the dairy herd and details concerning the sale of milk and cream. This information on dairy cattle is difficult to extract from the British census. Section nine is concerned with mechanisation on the farm and section ten with the application of chemical fertilisers and pesticides, listing the amounts used and the crops on which they have been used. In addition there is information on the amount of land that is held in the national soil bank, the acreage of strip cropping and contour ploughing and the acreage devoted to the federal crop diversification programme.

Section eleven deals with income and expenditure, including the amount spent on feeds, purchase of stock, seeds, fertiliser, fuel for the farm, machine hire and hired labour. Although this information is gathered only from a sample of farms it enables a detailed breakdown of cost to be made for the agriculture in a particular area. In addition information on the sources and amounts of income, on and off the farm, is included. Section thirteen concerns rental agreements of various types and the number of acres involved. If the farmer is a share cropper, details are entered in this section. The last section lists the saleable value of the farm and buildings, and the extent of the indebtedness of the farmer in terms of real estate. Clearly the census of agriculture in the United States is far more extensive than that held in the United Kingdom and involves the disclosure of considerably more detailed personal information, especially on matters of income and expenditure, than appears possible in Britain.

SWEDISH AGRICULTURAL DATA

Swedish agricultural data are available from a number of sources, the main one of which is the Farm Register which includes all holdings where agriculture is carried on. Data are collected on 26 June every year by the National Central Bureau of Statistics (SCB) and are published annually. This annual survey replaced, in 1968, a quinquennial agricultural census which was last held in 1966. The information for this register is collected by questionnaire, the first five sections of which list information about the farmer and his family and his level of education. Section six lists the number of hectares devoted to twenty-eight different arable crops and section seven the number of different types of livestock on the farm. Cattle and cows are subdivided into only three types: heifers, calves under one year old, and other cattle and cows. Section eight lists recent building activity on the farm and the next section deals with the employed labour on the farm, subdivided by age, sex and length of the employment period. Section ten lists the number of hectares devoted to fruit and vegetable crops and the number of square metres under glass. Finally details are collected on the number and size of different holdings owned or rented, including forest lands.

Considerably more detail concerning livestock, milk production and labour used on the farms is collected from a sample of farms every 5 March, and yield data are published for 420 yield survey districts. Finally data on cash incomes from agriculture are collected from 11,000 holdings using individual income tax returns to the SCB. Although data are scattered and many are available only on a sample basis, in total they provide a comprehensive and detailed set of information on all aspects of agriculture. This is all published for administrative areas; mainly the twenty-four counties but often for parishes. Many data are also available for sixty areas defined in accordance with natural farming conditions.

PROBLEMS IN THE USE OF AGRICULTURAL STATISTICS

The problems stemming from the use of agricultural censuses fall into two groups. First, there are those problems originating from the units of collection and/or the units of publication of the material and, secondly, there are those concerned with the reliability and comparability of the data both spatially and temporally. Jones[6] notes that one of the difficulties of using published sources of data is to find key maps showing the boundaries and the location of the units of area being used by the census. In the United States, although county maps are produced by the Bureau of the Census, they are not reproduced in the volumes of the agricultural census. In the United Kingdom, where agricultural parish summaries are available, maps of parish boundaries are not provided with the data. The more complex problems of the inappropriateness of the English parishes or of the counties of the United States as units for the presentation of agricultural material derive from the fact that the unit of operation in agriculture, and the unit for which original data are collected by the agricultural census, is the farm. But the confidentiality of these farm data are safeguarded in both the United States and the United Kingdom. Farms are grouped together and summary information is released only for these larger units.

The parish is unsuitable for this purpose. (The county in the United States has similar shortcomings, although less acute, in the areas of regular land-survey characteristic of much of the middle west.) The confidentiality of the individual returns is strictly adhered to (Section 80(a) of the United Kingdom Agriculture Act, 1947). The use of the parish as the unit for publication stems from a long tradition, especially in population enumeration. As early as the middle of the seventeenth century parish registers had been standardised by law so that they recorded all births, marriages and deaths. This was to become the basis of the population census which was first advocated in the British Parliament in 1753. From that date until 1801 the parish records remain the only source of population statistics, and when the full population census was started it was natural that the parish should continue as the basic unit for collection and tabulation of the material: it has continued so until the present day. This tradition was carried over to the agricultural census. Although the parish had little inherent advantage for the purpose, prior to the Census of Agriculture parish tithe records provide some information on local agriculture and the adoption of the parish as the basic unit for the census does provide an element of continuity.

The majority of the parishes in the United Kingdom date from before the tenth century AD, and the reasons for the particular boundaries are now often impossible to determine. However, it is clear that in many cases the boundaries were selected to contain a variety of the types of land in the area. Coppock[7] distinguishes two statistical problems stemming from the use of parish agricultural data which are also applicable to the use of any local government unit, for example, the United States' county: first, that the detail concerning the individual farms is lost and, secondly, that there is a range of variation about the mean parish data hidden within these parish summaries. The first of these is affected by parish size: the larger the parish the greater the degree of generalisation. English parishes vary greatly in size from a hundred acres or less to several thousand acres. This level of generalisation can be illustrated by an examination of the individual farms in comparison with the parish summary data (Figure 12). The variations in size of the statistical units can be equally well illustrated in the United States. Table 6 sets out some data for the largest and smallest counties in the state of Nebraska.

The range of variation within the parish or the county is partly a response to parish size and partly a response to farm size and diversity. This, in turn, will be influenced by the variety of relief, soil and drainage conditions found. Usually the smallest holdings show the greatest variation, so a parish of small holdings will have a considerable range of internal variation which will not be evident from the parish summaries.

Faced with these difficulties of using the parish units, it would seem that an improvement could be brought about by an amalgamation of parishes to create new units of approximately the same size. In practice, however, this is very difficult. Grouping is necessary because of the very juxtaposition of large and small parishes which makes amalgamation difficult. Small parishes must be amalgamated with neighbouring large parishes thereby making them larger still. This grouping is made more difficult when account is taken of the variety in relief, soil and other characteristics within the parishes which are being amalgamated. Situations

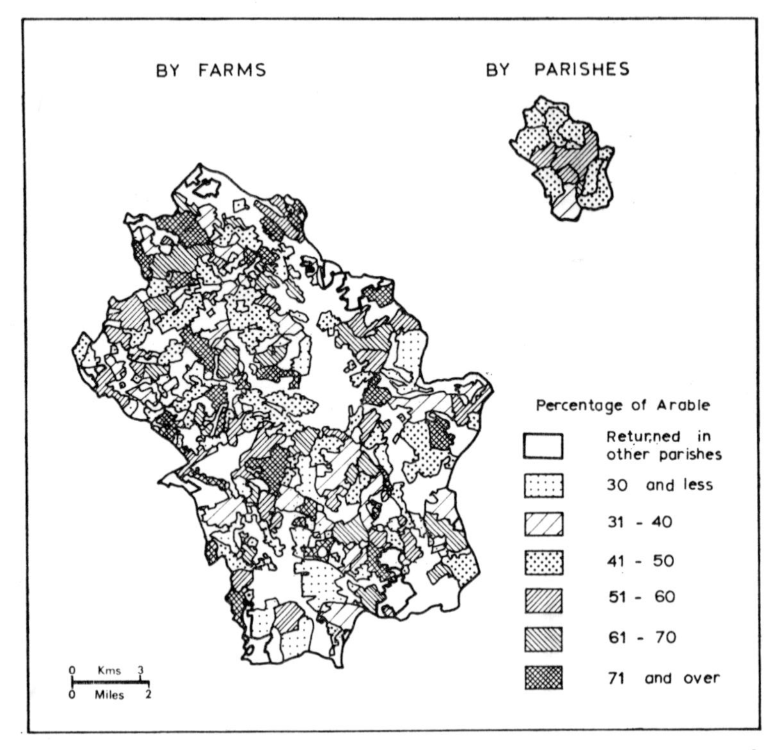

Fig 12. The differences between an agricultural and civil parish. The percentage of the land of each parish in arable cultivation is illustrated in the small diagram. The same data recorded for each farm in the area shows a much more confused pattern with very substantial areas recorded in parishes other than that in which they occur. (After Coppock, J. T., 'The parish as an agricultural unit', *Tijdschrift voor economische en sociale geografie*, 51 (1960), 317–26.)

where it is possible to interpret information about a single farm from the parish summaries as, for example, where there is only one farm in the parish, are avoided by amalgamation of that parish with a neighbour before release of statistics. The selection of this neighbour does not appear to be made on any statistical basis. On the whole any advantages of amalgamation to create more uniform statistical units are outweighed by the difficulties and by the loss of detail in the statistics which must result.

One further difficulty is what Coppock[8] calls the difference

TABLE 6 Number and size of farms in two counties in Nebraska, USA

	Cherry Co	*Johnson Co*
No of farms	686	850
Acreage	3,828,480	241,280
Acreage in farms	3,590,643	228,757
Average size of farm (acres)	5234·2	269·1

between agricultural and civil parishes. This difference stems from the listing of all of a farm in a single parish. The land of a farm may be divided among several parishes but has to be returned in one or other of them. With changing farm and parish boundaries this relationship is not static. The same type of problem is evident in the agricultural census in the United States. The land is returned under the county where the manager lives, or the county which has the farm headquarters, or, in any cases of remaining doubt, in the county which has the majority of the land of the farm. As a result some counties, especially in areas of large farms, have more land returned in farms than they actually possess: in 1964 forty-one counties had 50 per cent more land recorded than actually existed within their boundaries. In the Republic of Ireland the position is, if anything, more confusing. The livestock of a farm are returned for the parish in which the farmhouse lies, while the crops are recorded in the parish in which they are grown. A similar position holds in Sweden. This type of problem is reduced when the difference between the size of the farms and the size of the statistical units is great. It is less obvious, therefore, for all large statistical units. In conclusion, the use of administrative units for returning agricultural statistics is particularly unsatisfactory as they have no relationship with any of the facts of agricultural geography. Unfortunately there appears to be no suitable alternative at present.

In terms of the second group of problems, reliability and comparability of data, there are many shortcomings of all censuses of agriculture. For the census of England and Wales, Best and Coppock[9] point out some of these problems. Holding the census on 4 June each year in the United Kingdom raises its own problems. Spring crops, especially spring greens, may have already been harvested and the ground cleared, or even planted with a further crop: turnips and possibly sugar beet may not have been planted by this date. Land which is double-cropped is difficult to record accurately in a census held only once a year and this may lead to serious underestimates of certain crops; it may be particularly critical in horticulture. Also, in horticultural areas, where crops are interplanted, the acreage devoted to each type has to be estimated by the farmers. In the United States holding the census after the harvest, in November, avoids some of these difficulties, but in years of very late harvests some of the yields may have to be estimated by the farmers.

When considering the shortcomings of any census it is important to keep in mind the purpose for which the information is being collected; this may be very different from the purpose for which an independent research worker wishes to use the data. Horscroft defines the purpose of the census of England and Wales thus: 'the resulting information . . . indicates what measures are needed to achieve the objectives of agricultural and food policy'.[10] Yet it would be extremely valuable to be able to compare yield data as well as acreage data, and presumably this would be of considerable benefit when assessing future agricultural policy. Although the United States data on yields contains an element of assessment by the farmer and by the census office, none the less the census has a considerably enhanced value as a result.

Past errors in the English agricultural census can be traced to two sources: ignorance and ambiguity. Before the publication of large-scale Ordnance Survey maps many farmers must have been unaware of the exact size of their holdings and their fields. Also, through a lack of adequate records, farmers may still not be able to answer certain questions, especially with regard to numbers of young poultry. Ambiguity as a source of error has been largely eradicated over the years. It is still necessary to mention on the census form that hops are to be returned in terms of statute acres, not hop acres, and precise definitions are now given of that previously ambiguous term 'temporary grass'. Undoubtedly, some mistakes are still made but they are likely to be small ones.

LAND-USE SURVEYS

Another major source of data on agriculture is to be found in the land-use surveys which have been conducted in a variety of forms in most countries of the world. Land-use survey in the

United Kingdom grew up entirely independently of the agricultural census. The detailed investigation and mapping of the type of agricultural land use for each and every parcel of land in the United Kingdom can trace its history to the regional survey approach in geography which gained popularity in Britain at the beginning of the twentieth century. In this context the land-use survey became a medium for the accumulation of enormous quantities of data which could be best stored in the form of a map, but it involved little or no interpretation of this data. What little interpretation is to be found in these reports rests on a visual comparison of patterns laid side by side or as overlays.[11] A further line of development which encouraged the land-use survey was the growing importance of fieldwork in geography. Both fieldwork and regional survey, in so far as they contributed actively to the teaching of geography, were encouraged from an early date by the Geographical Association.

In 1930, under the guidance of L. D. Stamp,[12] the land utilisation survey of Britain was launched with some field trials to test a simple sixfold classification of land use into categories of arable; heath, moor and rough pasture; orchards and nursery gardens; meadowland and permanent grass; forest and woodland and urban areas. This classification had to remain simple because the field force, to be mobilised to gather the data, was to consist of approximately 20,000 largely untrained school children and university students. The field work for the survey was completed on six-inch-to-the-mile maps and the information on these was then transferred to one-inch-to-the-mile maps for publication. Originally it was intended that there should be reports to accompany each of the published maps but this was later contracted to the preparation and publication of a report for each county. These reports were published between 1936 and 1948, the length of this publication period being unavoidable owing to the war.

These county reports followed the general lines established by the earlier regional surveys and concentrated on the relationships between the physical environment and agricultural land use, saying little about the human decisions and economic considerations which had influenced the land-use patterns. The reports contained little theory; they were descriptive and, in common with the whole survey, form what is best described as a historical document: a representation of the land use at a particular period of agricultural development which can then be compared with other such periods. Such comparisons are limited by the time taken for the completion of the survey so that only long-term changes can be included, though short-term fluctuations would have existed within the survey time.[13]

Although the first land-use survey of the United Kingdom was primarily an academic exercise, the idea was adopted later, using the survey as a data base for decision-making concerning matters of land use.[14] The country is small enough to have made a complete survey feasible and the growth of the regional survey in geography in the United Kingdom made such a survey practicable at the time.

The second land-use survey of the United Kingdom, launched in 1960 by Alice Coleman,[15] was conceived largely as a necessary exercise to bring Stamp's survey up to date, providing further information for comparative studies, and in response to academic curiosity as to the nature of changes since the last survey. The survey has been conducted with a more detailed classification,* and the maps are published at the 1:25,000 scale. By mid-1972 114 of these maps had been published and all the field survey sheets for England and Wales were complete and available for consultation. In addition all land use has been placed into one of five basic types of environment (wildscape, farmscape, townscape, marginal fringe and urban fringe) and maps of these categories are shortly due to be published at the scale of 1 : 400,000.

In 1949, reflecting an increasing concern for the food resources of the world, and as a result of the success of the first British

* The classification is:

Arable: cereals, ley legumes, roots, green fodder, industrial crops, fallow.
Market gardening: field vegetables, mixed market gardening, nurseries, allotments, flowers, soft fruit, hops.
Orchards: with grass, with arable, with market gardening.
Woodlands: deciduous, coniferous, mixed, coppice, coppice with standards, woodland scrub.
Water and marsh: water, freshwater marsh, saltwater marsh.
Heath, moorland and rough land: fifteen vegetation types overprinted.
Settlement: commercial and residential, caravan sites.
Unvegetated.
Open space: tended but unproductive land.
Industry: manufacturing (sixteen categories overprinted), extractive, tips, public utilities.
Transport: port areas, airfields.
Derelict land.

land-use survey, with Stamp's guidance the International Geographical Congress proposed a world land-use survey. This ambitious project has had limited success because the techniques adopted for a country the size of the United Kingdom, with the degree of small-scale complexity of land use, are not likely to be applicable on a world basis. There are three specific difficulties: first, a worldwide land-use classification was very cumbersome and in practice extremely difficult to devise. Secondly, over much of the world topographic maps are of poor quality and small scale in contrast to those available in Britain: most of these maps are quite unsuitable for detailed field mapping. Thirdly, there was lacking the vast army of volunteer field workers that made Stamp's survey of Britain so successful: clearly for any country considerably larger than the United Kingdom the survey team would have to be proportionately larger. Cyprus, the Sudan and Iraq were surveyed under the auspices of the world land-use survey[16] but many countries, especially in Europe, continued independently with their own surveys. For example, the Italian National Research Council and the Director of the Census (Consiglio Nationale delle Ricerche; Direzione Generale del Catasto) have produced land utilisation maps for the whole of mainland Italy, Sicily and Sardinia in twenty-six sheets to a scale of 1 : 200,000. These sheets are published by the Italian Touring Club of Milan. The classification in twenty-one* categories naturally reflects agriculture typical of the Mediterranean.

The conditions making the land-use survey of the United Kingdom so successful were, to a considerable extent, unique. One would not expect to find work in the United States following similar lines. However, interest in field mapping of land use by geographers in the United States can be traced to a number of

* The classification is:

Arable: dry arable, dry arable with trees, irrigated arable, irrigated arable with trees, rice, market gardening.

Tree crops: vines, olives, olives and vines mixed, oranges and lemons, apples, pears etc, nuts (subdivided into almonds, hazel nuts and carob indicated by overprinting).

Woodland: deciduous, coniferous, mixed, chestnuts.

Grass: dry meadow and meadow with trees, irrigated meadow and meadow with trees, uncultivated meadow and meadow partly or temporarily arable.

Vacant land.

Urban land and other forms of use.

geographers at the University of Chicago in the years following 1915.[17]

Categories of land use were defined and methods of intensive field survey indicated, though essentially as an exercise in field work rather than as an analysis of agriculture. Interest in 'local studies' was keen, especially in the mid-western states between 1915 and 1925. Several similarities and differences exist between this early work in the United States and later work in the United Kingdom. In the United States, there was never any desire or attempt to survey the whole country. In fact, the emphasis was on local study areas. The impracticability of a full land-use survey in a country the size of the United States has meant that no full survey has ever been attempted. Secondly, the early United States work was usually directed at hypothesis testing.[18] This was in marked contrast to the accumulation of facts felt to be a prerequisite of hypothesis generation which characterised the later approaches in the United Kingdom. Despite these fundamental differences there was, in both countries, a concern for the role of field teaching which became such a fundamental part of geography.

SAMPLING AND LAND-USE DATA

The differing history and purpose of land-use surveys in the United States as contrasted with the United Kingdom meant that geographers in the United States were the first to adopt sampling procedures to eliminate the necessity of comprehensive field mapping which, apart from the considerable investment of time and effort needed for its completion, can easily become an end in itself and detract from the purpose of collecting the information.[19] One of the earliest applications of sampling to gather land-use data is provided by the geologist Trefethan[20] who tested the use of traverse lines at right angles to each other. He found that this considerably speeded up the process of acquiring land-use statistics without the necessity of a complete survey. Of course, no sample survey can give complete information about any particular plot of land unless it happens to fall within one of the sample areas, but the sample data can be used to estimate the percentage of different land-use classes in the area sampled.

Since 1936 there have been a number of attempts at the application of line and area sampling techniques, of various degrees of sophistication, to land-use data.[21] The most complex is known as stratified systematic unaligned sampling.[22] The actual method of this type of sampling has been described elsewhere.[23] It is necessary to impart a degree of randomness into a systematic sample where the sample points are located at the intersections of a regular grid. The major shortcoming of this form of sample is that the points, being regularly distributed, may coincide with some regularity in such surface features as, for example, parallel ridges and valleys. All the points may fall either in the valleys or on the ridges, thereby giving rise to considerable sample bias. Stratified systematic unaligned sampling permits one sample point to fall in each grid square and a more even distribution of points is achieved than by purely random methods, but the location within each square has a random element (Figure 13). The resulting sample contains many of the advantages of randomisation and stratification with the useful aspects of systematic sampling, while avoiding possibilities of bias because of the presence of

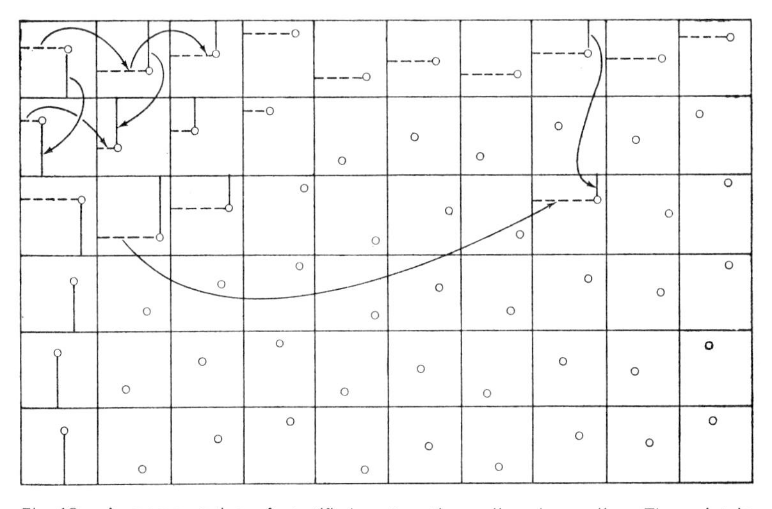

Fig 13. A representation of stratified systematic unaligned sampling. The point in the top left-hand square is first located by random numbers in a north-south direction and in an east-west direction. For the square immediately to the right the same east-west number is used with a new north-south number. This process is continued across the row. For the square below the first the same north-south number is used with a new east-west number. This process is continued down the first column. The point in the square diagonally south-east of the first square takes a north-south value from the square above and an east-west value from the square to the left. This process continues until all the squares have points within them.

periodicities. This type of sample has been shown to be the most accurate of all sampling systems for the estimation of land-use data.[24] Before this system was developed there were other authors who suggested that sampling areal data such as land use is less successful using point sampling than by a system of transect samples.[25]

By sampling, and thereby reducing the time and effort of the full land-use survey, three major criticisms can be at least partly avoided.

1 The duration of the data collection stage is shortened and the survey approximates more closely to the census ideal of recording the state of land use at a particular moment in time.
2 Basic details of the land-use patterns can be recorded more quickly than through full land-use surveys.
3 In contrast to the full land-use survey which may have positively harmful effects in terms of teaching, the sampling method not only involves students in field survey techniques but has the additional benefit of familiarising them with some of the principles of sampling.[26]

In many countries of the world because of size, lack of topographic maps and lack of the necessary skilled labour force, the use of some sampling technique may not be an alternative but a necessity if a land-use survey is to be conducted.

SAMPLE FARM STUDIES

In cases where the parish is very unsatisfactory as a data unit it may be possible, or indeed necessary, to examine the farms themselves. Agricultural economists have been leaders in the use of sample farm data. Although not primarily interested in land use it has been possible in these surveys to gather detailed information about the areal distribution of the different types of the operating units of agriculture.[27] Among the few geographers to have used this technique, J. W. Birch[28] studied the farming patterns in the Isle of Man.[29] He found, with only a few exceptions, the parishes of the Isle of Man far from internally homogeneous and this led him to reject the use of parish agricultural statistics as an appropriate source of data for a study of the farm structure of the island.

The use of individual farm statistics presents two difficulties, the first of which is the very large number of farms that may be involved, and the second the confidentiality of the 4 June government returns. These difficulties have meant that most attempts to use farm data on an extensive scale have been confined to official bodies.[30] The answer to these two difficulties has been for research workers to consider only a sample of farms and to gather data from these farms by a questionnaire of their own. Without a list of all the farms in the study area it is difficult to remove bias from the method by which the farms are sampled. Farms have two characteristics which have an areal distribution and which can be used as a basis for sampling. These are the location of the farmhouse or the centre of farming operations and the actual land area of the farm. Using random numbers to generate random grid references it is possible to select a number of randomly distributed points throughout the study area. All or most of these will fall within the land belonging to one farm. These farms could be used as the sample. But, if the area had within it a few large farms, the chances of the majority of these sample points falling within the area of the large farms is very high. The final sample will, almost certainly, be biased in favour of the large farms which, although few in number, cover a substantial proportion of the land area. Further, to select the nearest farmhouse to the random points might also bias the sample towards the large farms. Dispersed large farms, separated from their nearest neighbours by considerable distances, have a greater chance themselves of being the nearest neighbour to the random points than have the small farms with only a short distance to nearest neighbours. Any form of areal sample will not give an unbiased representation of all farm sizes: the chance of a farm's selection will be proportional to the size of the farm.[31]

Some improvement might be brought about by selecting the nearest neighbour to each selected farm, assuming that the different sizes of farms were thoroughly intermixed. If there is, within the study area, a part which has predominantly small farms and a part which does not, it would be advisable to stratify the sample and to select farms at random from each part of the study area separately.

If what is desired is a representative cross-section of the farming structure in the area, in the last analysis there is no alternative to listing the farms and selecting from this list at random. The identification and listing of the farms is a tedious, and in some cases virtually impossible, job; even the Ministry of Agriculture does not know the exact location of all the farms from which it receives information at least twice a year. Given the difficulties of creating a suitable sample design, this system is not likely to be of much use in countries where other forms of agricultural data are not available. It is often necessary to resort to some second-best alternative.

However, once the sample of farms is selected it can be used to examine not only land-use patterns but also the entire structure of the farming operation in the area. No matter what the difficulties, an attempt can be made to obtain a statistical sample of farms in such a way that reference can be made to the population of farms from which the sample has been drawn. In the case of other methods of selecting 'representative' farms this is unlikely to be valid.[32]

REMOTE SENSING OF LAND USE

The difficulties created for land-use survey by a lack of topographic maps and a suitable labour force can perhaps be better solved by the use of one or more of a range of remote-sensing techniques, some of which have been available and utilised for many years, some of which date only from recent military developments. Remote sensors are data-gathering devices which operate without being in direct contact with the object about which they are collecting information: the human eye is a remote sensor, as is the camera. In normal usage, aerial photography is considered sufficiently remote to be included under this heading, but ground-level photography is not.

The earth's surface receives radiant energy from the sun. The wavelength of emitted radiation from a body varies inversely with the temperature of that body, the radiation from the sun lying within the wavelengths 0·2 and 4·0 microns. Radiation from the earth, being a cooler body, lies mainly between 4·0 and 50 microns. There are, therefore, two types of radiation which can be detected above the earth's surface. First, a proportion of the sun's radiation is reflected from the earth's surface back into the atmosphere. The relative absorption of the surface determines the character of the remaining reflected radiation which can be detected above the earth. For example, plants look green because the chlorophyll in the plant absorbs all the visible light from the sun with the exception of the green part of the spectrum; this is then reflected back into the atmosphere. Secondly, the earth's surface absorbs the sun's radiation and re-radiates it back as radiant energy. The degree to which an object is able to emit such radiation is called its emissivity. This radiation from the earth's surface enters the atmosphere, where the carbon dioxide and other constituents absorb most of it. There are three 'windows' in this absorption which allow wavelengths of 3–5 microns, 8–14 microns and 16–24 microns to be detected with sensing devices. These three windows are extensively used in remote sensing.

The electromagnetic spectrum is shown in Figure 14, the portions used in remote sensing being divisible into four sections:

1 Visible light (0·4–0·7 microns); we are interested in those portions of this waveband which are reflected from the earth's surface.
2 Reflected infra-red (0·7–0·9 microns): otherwise called near-infra-red or photographic-infra-red, and photographing this radiation is often called false-colour photography. This is radiation which is reflected from the surface of the earth but which is of a longer wavelength than visible light.
3 Thermal or emissive infra-red (1·0–1,000 microns): radiation which is absorbed by an object at the surface which then re-radiates the energy. Because the object is substantially cooler than the sun, the wavelength of the radiation is longer. It has to be collected by a sensor specifically designed to detect radiation in this waveband as the energy can neither be seen nor does it affect photographic emulsions. Variation in emissive-infra-red radiation is mostly the result of variations in surface temperature.
4 Microwaves (1,000+microns): sensors for the shorter wavebands are passive, that is, they merely record the radiation they receive. Use is made of microwave radiation by actually generating it and bouncing it off objects on the surface of the earth. The proportion of the radiation which is reflected is then recorded by the sensors. This is the principle of radar and it is some of the radar wavebands which are used for this type of remote sensing. Variations in the surface will lead to variations in the strength of the reflected signal in just the same way as variations in the surface affect the level of reflected visible radiation from the surface.

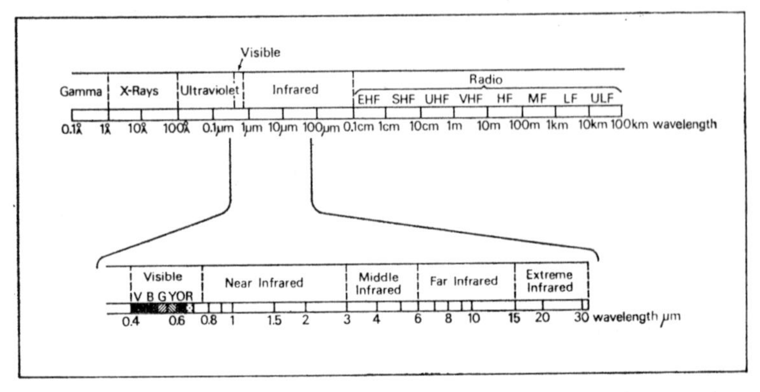

Fig 14. The electromagnetic spectrum showing the relationship between those wavelengths used in remote sensing and the remainder of the spectrum.

VISIBLE RADIATION

Board admits that 'technological developments have moved so fast that land-use surveys as we know them are now out of date as a method of amassing spatial information on the way the land is used'.[33] The first of these technological developments which went some way towards replacing the ground survey came with the extensive application of vertical air photograph interpretation, a technique which rapidly developed from wartime aerial reconnaissance. The complete coverage* of the United Kingdom in 1946–7 at a scale of 1 : 10,000 by the Royal Air Force provides a fund of information which is used considerably more extensively than the first land-use survey. On the other hand, the uses to which this type of black and white photography can be put are limited by the extent to which features of agricultural land use can be detected by visual interpretation of the various shades of grey between black and white.[34] The recognition of classes of land use must remain necessarily crude.

With the advent of cheaper colour photography it has been possible to identify more accurately different classes of land use and the latest developments have used the reflectance of different

* Areas of military significance were excluded from this survey and photographs of selected sites are not available for reproduction. As these sites are normally military airports this does not provide a significant limitation on the use of this survey for the study of agriculture.

types of surface in narrow bands within the visible spectrum. Purdue University Laboratory for Agricultural Remote Sensing used twelve waveband divisions between 0·4 and 1·0 microns, the last two divisions of which were in the infra-red part of the spectrum. Pestrong[35] found that the wavelengths between 0·6 and 0·63 microns (the orange part of the visible spectrum) were best for detecting bottom features of channels in saltmarsh areas, although the effectiveness of this waveband depends on the degree of turbidity of the water. In general Pestrong concludes that:

> A colour reproduction is superior to one in black and white because it looks more like the original scene with many of the subtle shadings of hues and saturations represented. Colour photographs, therefore, provide the interpreter with an extra, important factor, in addition to the form and shape parameters normally associated with image interpretation;

and further:

> if economic considerations dictate the utility of only a single type of photograph imagery, for general field interpretation, the image most useful would be colour positive transparencies.

Because of its obvious advantages over black and white alone, colour photography has been extensively applied to land-use survey work.[36]

REFLECTIVE INFRA-RED

Reflected infra-red radiation has a longer wavelength than visible light but can still be detected on such photographic emulsions, especially designed for the purpose, as Ektachrome Infrared which is sensitive to the infra-red, green and red wavelengths. This film is normally used in conjunction with a yellow or orange filter to remove blue light to which all layers of the photographic emulsion are sensitive. On the resulting positive, objects reflecting predominantly green light show as blue, objects reflecting red show as green and objects which reflect a large proportion of the received infra-red radiation show as red. Infra-red film was first used extensively for camouflage detection during World War II.[37] Vehicles painted green to merge with the surrounding vegetation could be detected easily as they became blue on this film while the surrounding vegetation, with a high infra-red reflectance, showed as red or red-blue.

Infra-red imagery is of particular use for the remote sensing of vegetation because of the characteristic of plants in reflecting a high proportion of the received infra-red radiation. Different types of plant reflect at different intensities as a result of the different cell structure of the leaves. The most significant difference is between broad-leaved and needle-leaved trees. The degree of reflectance is the result of the structure of the parenchyma cells in the mesophyll layer* of the leaf. In broad-leaved vegetation these cells are widely spaced with considerable voids between them. The radiation in the short infra-red range (0·7–1·2 microns) enters the leaves and is reflected from cell to cell through these voids. In the process much is 'bounced' out of the leaf and can therefore be detected by remote sensors. In needle-leafed trees these parenchyma cells are more densely packed, with few voids, and the infra-red radiation is absorbed and reflectance substantially reduced.[38] In certain types of plant disease the voids in the mesophyll layer become filled with fungus, cutting down the amount of reflected infra-red radiation. This enables foresters to detect tree disease before it manifests itself in the outward appearance of the trees. Cooke and Harris[39] conclude that 'a pre-eminent geographical application of false colour (reflected infra-red) photography lies, therefore, in vegetation studies and surveys of agricultural land use'. They found in a study of the Isle of Man that it was possible to classify accurately rough grass, dormant bracken, growing gorse, areas recently burnt and individual tree species in the wooded areas from reflective infra-red imagery.

It is important to be clear as to the exact nature of this radiation. Fritz[40] discussing Ektachrome Infra-red Aero Film, makes the position quite clear:

> There is a tendency to equate infra-red radiation with heat and to expect that this film can be used to record temperature differences. Tests have shown that, if there is no other source of radiation to affect the film, an object heated to 650°F

* The mesophyll layer of a leaf is the spongy material just below the surface. This is made up of parenchyma cells with quite large intercellular spaces.

will just be recorded by the infra-red sensitive layer of the film exposed for fifteen minutes at F 2·0. The film will not record temperature differences at our environmental temperatures.

To record surface temperature differences we need to record emissive infra-red radiation at considerably longer wavelengths which does not affect ordinary photographic emulsions.

MULTI-WAVEBAND PHOTOGRAPHY

Because different sections of the reflected electromagnetic spectrum are useful for different purposes in remote sensing, it is often convenient to be able to take several vertical photographs, within limited wavebands, simultaneously. For this purpose it is usual to employ a multi-band camera taking nine simultaneous frames. Six frames operate within the visible spectrum and three in the reflected infra-red. The sequence used by Pestrong[41] and Cooke and Harris[42] is shown in Table 7. By using all these images together an accurate picture of the land-use and vegetation pattern is obtainable. However, even more sophisticated techniques are available.

TABLE 7 Multi-band camera configuration

BAND 1	BAND 2	BAND 3
blue	blue-green	green
0·4–0·44	0·46–0·5	0·525–0·55
BAND 4	BAND 5	BAND 6
green-yellow	orange	red
0·55–0·59	0·6–0·63	0·65–0·71
BAND 7	BAND 8	BAND 9
	reflected infra-red imagery	
0·7	0·75	0·825

If different types of plants are detectable using different wavelengths then each type of plant and each type of land use will have a 'signature' or pattern of reflectance over a range of wavelengths which will allow it to be identified.[43] By plotting percentage reflectance against wavelength this signature can be established. Figure 15 shows the characteristic reflectance signatures of broad-leaved and needle-leaved trees through wavelengths 0·4–0·9 microns. The clear separation in the infra-red bands shows

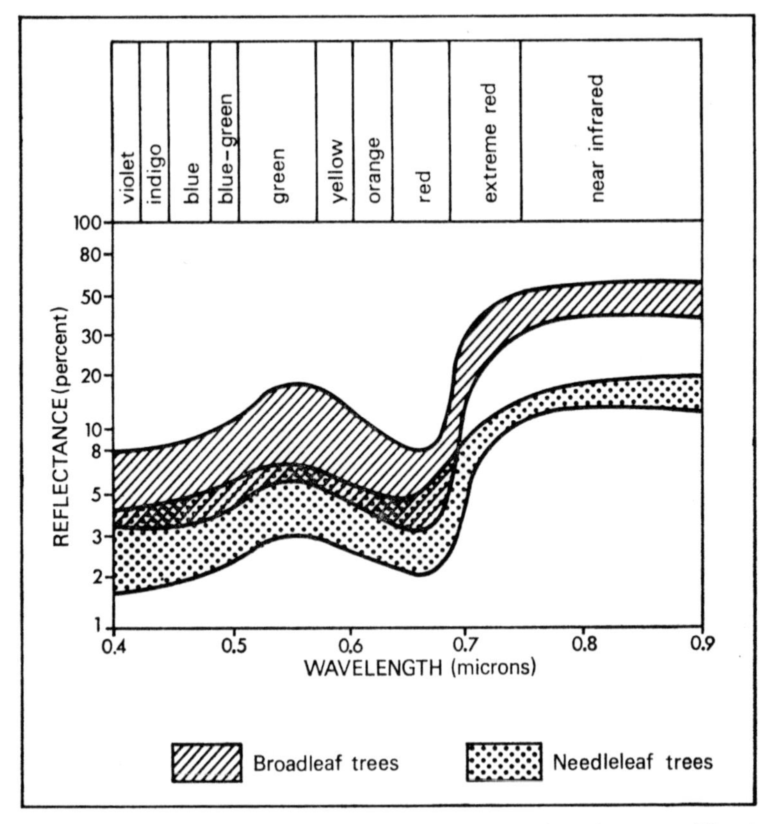

Fig 15. The percentage reflectance of broad- and needle-leaved trees at different spectral frequencies. The 'signatures' become clearly separated in the infra-red section of the spectrum.

how these two major vegetation types can be separated. Figure 16 shows other characteristic signatures of different land surfaces. Unfortunately the form of this signature is affected by factors other than the nature of the plant itself. The moisture content of the leaf is, for example, a major factor. Water absorbs some of the radiation, especially above the 1·0 microns wavelength, so the greater the water content of the leaf the less will be the reflectance of the longer wavelengths (Figure 17). Similarly the season affects the amount of chlorophyll present in the plant and this also affects the degree of absorption and reflectance. Lastly, the relative amount of sun and shade on the leaf's surface affects the reflectance levels.

These problems serve to emphasise the importance of establishing good 'ground truth'. In this case this means correctly establishing the characteristic signatures of different types of land use to which the recorded ones can be compared.

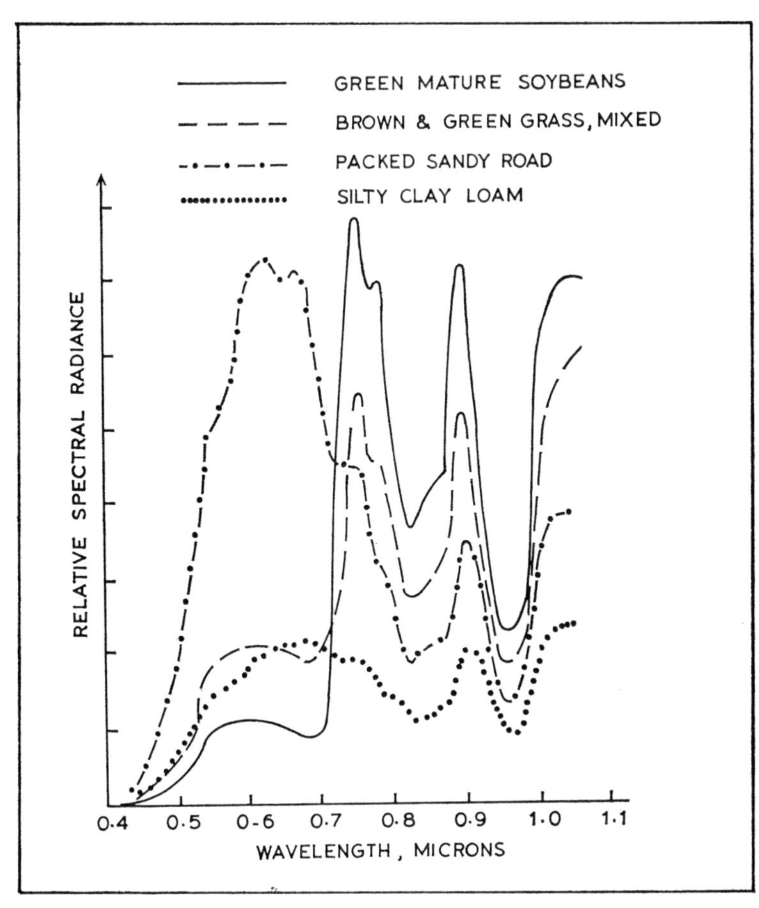

Fig 16. Reflectance of different crops and ground surfaces at different spectral wavelengths. (After Purdue University, *Laboratory for agricultural remote sensing, 3, Annual Report, 1968.)*

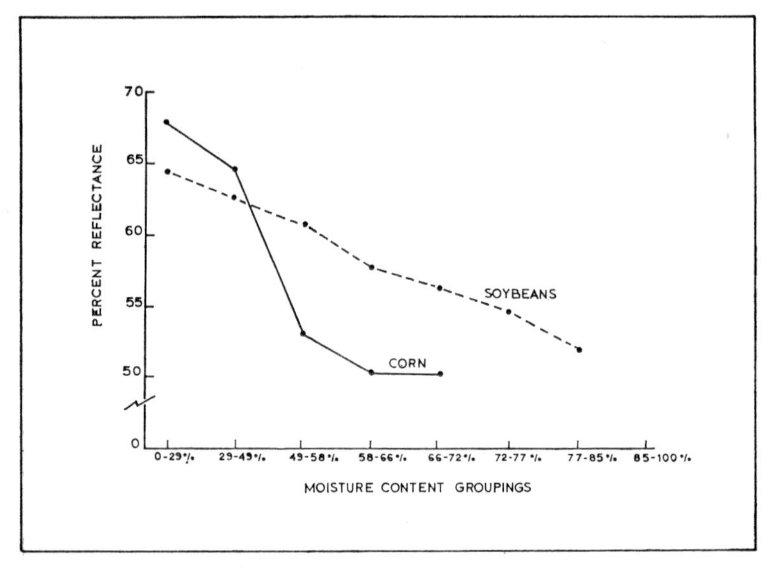

Fig 17. A comparison of the reflectance at 1·05 μ of corn and soybeans under different moisture content conditions. (After Purdue University, *Laboratory for agricultural remote sensing, 3, Annual Report, 1968.)*

Despite these difficulties it has been shown[44] that the reflectance pattern of different crops can be used to classify land use with considerable accuracy. Characteristic patterns can be matched by computers to actual patterns and each of a series of sample locations can be placed in its correct category. In a trial set of samples a very high degree of correspondence was found between type signatures and the recorded signatures (Table 8).

THERMAL INFRA-RED

Radiation of wavelength longer than 3·0 microns is the emissive infra-red band. This radiation is longer than that received from the sun and so is longer than can be reflected from the earth's surface. Energy from the sun is absorbed and re-emitted in the form of radiation of longer wavelength. Emitted radiant energy from the earth reaches a peak at 9·7 microns. Three 'windows' between 3–5, 8–14 and 16–24 microns exist for this emitted radiation. In the wavebands between these windows the radiation is

Class or cover type	Field number	Percentage correct classification	Total no. of RSUs*	Number of RSUs classified into: Bare soil	Wheat	Oats	Rye	Red clover	Alfalfa	Corn	Soybeans
Bare soil	36–3	100	56	56	0	0	0	0	0	0	0
Wheat	31–12a	88·0	100	0	88	12	0	0	0	0	0
Wheat	31–12b	99·7	336	0	335	0	0	0	0	0	0
Wheat	6–1a	100	138	0	138	0	0	0	0	0	0
Wheat	6–1b	100	210	0	210	0	0	0	0	0	0
Wheat	6–14	99·4	506	0	503	3	0	0	0	0	0
Wheat	12–10	26·5	520	0	138	36	186	0	0	0	0
Oats	1–11	73·9	460	0	2	340	0	32	0	51	34
Rye	31–18	73·0	126	0	2	2	92	0	0	1	8
Red clover	1–1	81·8	336	0	0	3	0	275	11	43	0
Red clover	6–7	72·4	504	0	3	80	0	365	7	46	3
Red clover	7–28	100	174	0	0	0	0	174	0	0	0
Red clover	12–8	83·6	500	0	0	4	0	418	54	1	0
Red clover + alfalfa	7–25	48·5†	468	0	1	183	1	147	80	25	0
Alfalfa	7–22	52·4	418	0	0	21	0	100	219	10	6
Corn	36–4	99.5	560	0	0	0	0	0	0	557	3
Corn	36–8	98·1	160	0	0	0	0	1	0	157	2
Corn	1–15	75·6	352	0	0	0	0	0	0	266	81
Corn	1–14	49·2	132	0	0	0	0	0	0	65	67
Corn	1–9	79·2	168	0	0	0	0	0	0	133	34
Corn	1–3	34·2	351	0	0	0	0	0	0	120	231
Corn	1–4	8·8	297	0	0	0	4	2	0	26	259
Soybeans	30–4	68·9	273	0	0	0	0	3	0	82	188
Soybeans	31–1	89·2	185	1	2	2	9	0	1	0	165
Soybeans	36–2a	98·8	168	0	0	0	0	0	0	2	166
Soybeans	36–2b	97·0	168	0	0	0	0	0	0	5	163
Soybeans	31–14+15	14·5	585	0	0	0	0	322	34	104	85
Soybeans	36–7	85·8	650	0	0	0	0	0	0	92	558
Soybeans	1–12	41·1	336	0	1	0	0	0	0	197	138
Soybeans	6–16	96·0	125	0	0	1	2	0	0	2	120
Soybeans	12–2	94·6	168	0	0	0	0	0	0	9	159
Soybeans	12–3	88·4	1,833	10	0	0	0	0	0	66	1,620
Soybeans	7–23	94·2	481	5	0	0	0	0	0	22	453

* RSU is the sampling unit within each field.

† 48·5% correct classification represents the combination of RSUs classified as red clover plus those classified as alfalfa. Red clover alone had a 31·4% correct classification.

(After Laboratory for agricultural remote sensing, 3 (1968), Annual report (Agricultural experimental station, Purdue University, Lafayette, USA).)

absorbed by the carbon dioxide, water vapour and other constituents of the atmosphere. Of these windows that between 8 and 14 microns is known as thermal infra-red radiation. Different radiation levels are the result of differences in the surface temperature of objects and the differences in their degree of emissivity.

Unlike the reflected infra-red radiation, which can be imaged directly on infra-red sensitive film, some type of transformation is needed in the case of thermal infra-red. A method often used and described by Sabins[45] is to use germanium wire coated with copper or mercury: this makes a good photo-conductive detector. Its electrical conductivity changes as a function of the infra-red radiation striking it. As a result the infra-red energy is converted into a variable electrical current which can be directly recorded on magnetic tape or used to illuminate a glow lamp. The greater the infra-red radiation the brighter the lamp. This variable brightness can then be recorded on photographic film.

In operation, the infra-red sensor is 'scanned' across a strip of country at right angles to the flight path of the aircraft holding the sensor with a parabolic mirror used to focus the energy on to the detector. Each scan of the sensor adds a strip to the picture on the photographic plate until the entire picture is built up. As a result this technique is known as infra-red linescan (IRLS). The density of these pictures can be measured, and, if necessary, contoured, by using a microdensitometer.[46]

Surface temperature sensitivity of this equipment can be as low as 0·1 °C. Light areas on the photograph are relatively warm because the greater level of emission is transformed into greater light in the glow lamp. Although most applicable in the field of geological interpretation, thermal infra-red emission has applications in agriculture. For example, Sabins[47] shows how irrigated fields which are flooded can be distinguished from irrigated fields which have no surface water. The flooded fields are relatively warm while the irrigated fields which have no surface water image as cold because of surface cooling through evaporation. Such small variations in surface temperature as those found in exposed and sheltered sides of hedgerows and recently ploughed land can also be distinguished.[48] Clear, calm conditions are required and results can be quite badly affected by humidity, high winds or air pollution. In general this technique has more application in the field of terrain studies than in land-use studies.

MICROWAVE SENSING

Wavelengths longer than the infra-red are not emitted or reflected naturally from the earth's surface. The sensing devices using microwave wavelengths, therefore, have to be active systems having their own energy source. The systems using radar wavebands, for example, generate and emit the energy, and the sensors detect the level of reflected radiation; variations in the proportion of the emitted radiation which is reflected back to the sensors are the result of variations in the nature of the ground surface. Figure 18 shows how this proportion varies for a number of different surface types.

There are technical difficulties with this system, especially the 'shadow' directly under the aircraft which extends to 45° on either side of the vertical. On the other hand, as it generates its own energy, the advantages of the system are that it is independent of atmospheric conditions and can be used successfully at night. It can be used in places where cloud cover is habitually great and all other forms of sensing are unsuccessful. Because of the range of the sensors, the surveys can be flown at considerable heights and are therefore very economical in their use of film. This technique may have considerable application in under-developed countries where the resources for such surveys may be limited. Interpretation is not easy, especially in distinguishing between 'shadows' of objects and surfaces which are genuinely low in their reflecting qualities, both of which will appear similar in the resulting photographic plate. Because the radiation is emitted on either side of the plane the technique is known as SLAR (Sideways Looking Aerial Radar).

CONCLUSION

With all remote sensing techniques information about the areas surveyed, known as ground truth, is essential for correct interpretation. This is obviously necessary to be able to know what the recorded image implies. As a result it seems unlikely that land-use surveys will ever be completely independent of ground surveys. Crop signatures change continuously with the seasons and with the local climatic conditions and soil properties; thus, in order to establish the 'characteristic' signatures for computer matching, ground surveys are essential. Future developments in land-use survey would appear to lie in some combination of sample

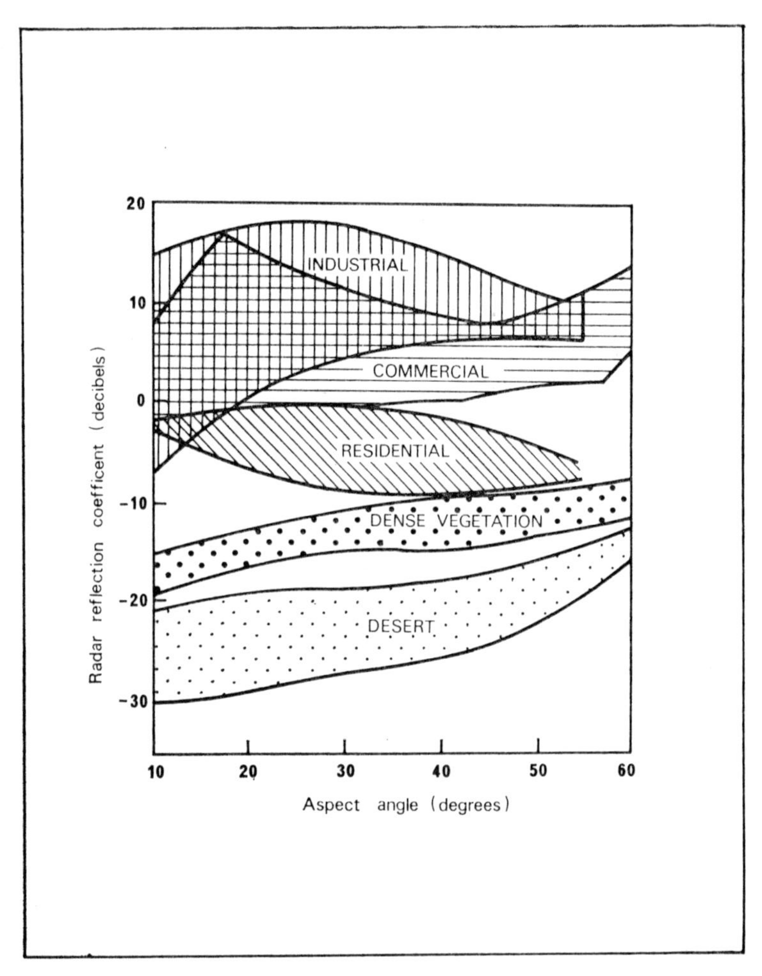

Fig 18. The relationships between the amount of radar pulse reflectance, the aspect angle of the pulse, and different types of land surface. (After Newbry, L. E., 'Terrain radar reflectance study', *Photogrammetric engineering*, 26 (1960), 630–7.)

ground studies and more extensive studies based on aerial coverage with multi-band imagery. Board[49] also concludes that ground surveys of sample areas and detailed areas where there are considerable local-scale variations in land use will still be needed. Land-use survey as a field method in geography will also undoubtedly be needed for a considerable time. The ultimate development for remote sensing is to use satellites to provide platforms for the various sensors. This proposal, first mooted in 1964,[50] was brought to fruition with the initiation of the earth resources technology satellite programme (ERTS). The resolution of the imagery from satellites so far released is of little use for detailed agricultural studies. In fact, despite a number of publications of the photographs derived from the Gemini series of satellite flights[51] there has been little published using satellite data, except in meteorology in which the application of space photography is both obvious and widely utilised. The continuance of the ERTS programme should make considerable progress in this field of more 'remote' remote sensing.

CHAPTER FOUR

Regionalisation and Classification

THE REGION IN GEOGRAPHY

THE use of the regional method by geographers has taken several contrasting forms. Despite, and to some extent because of, these contrasting uses, the regional concept has been a central part of geographical studies for at least a hundred years. In all cases the term region is distinguished from area by some assumption of homogeneity, not necessarily over the whole of a recognised region, but at least over a substantial part of its core area.

It is not necessary here to recount the history of regional thinking in geography[1] but it is important for much of the further work on agricultural regions to make the distinction discussed by Whittlesey[2] between single-feature, multi-feature and total regions. The approach to single-feature regions is to map areas of recognised homogeneity according to single features—eg slope, soil, rainfall or corn acreage—and to compare the regional boundaries so formed with those created by using other criteria by some process of map matching. Whittlesey recognised three subdivisions within the category of multi-feature regions:

1 those where highly cohesive factors have been used; for example, factors derived from the same process, such as soil regions defined from slope, soil and drainage characteristics, or farming regions defined from an examination of crops and livestock;

2 those which have used factors derived from different processes; for example, economic regions defined from an inclusion of agricultural and industrial variables;
3 those defined from an association of factors which are hardly connected; for example, natural regions defined from an association of all possible physical factors of the environment, or human regions defined from an inclusion of all factors relating to man.

Total regions should perhaps be a fourth category of multi-feature regions as they are defined taking all factors of the environment together. A total region is an area of the earth which has some unique combination of all factors such as East Anglia, the Mid-West, or, on a wider scale, the Middle East.

There is a further generic division in regional thinking of some significance to agricultural regions: this is the distinction between uniform and nodal regions. Nodal regions are uniform in terms of some aspect of their organisation or structure: for example, dairy regions could be produced by delimiting milk collection and marketing areas. If we add to our consideration other types of agricultural marketing and distribution we create a multi-feature nodal region. In contrast, a uniform region approaches uniformity in terms of criteria, not of organisation. An agricultural region based on a recognised uniformity of agricultural production would be such a uniform region. This is the approach most often used in agricultural geography.

REGIONALISATION AS CLASSIFICATION

Bunge[3] develops the thesis that there is no essential difference between the operations of classification and regionalisation. All sciences have, in their history, a classificatory or taxonomic phase and geography is no exception. The geographer's variables, on which he divides his data, are spatial variables and therefore the classes he produces from these variables have a spatial extent and become regions or areal classes. Thus the class of climate called Mediterranean is used to locate and define Mediterranean-type climatic regions. Bunge takes this argument further:

> In fact there appears to be a complete isomorphism. A single feature region is an areal class which is based on a single differentiating characteristic. A multiple feature region is an

> areal class based on more than one differentiating characteristic. A particular 'compage' translates into an areal class using many, but obviously a finite number, of differentiating characteristics.[4]

There remain two inter-related features of a region which make it rather a special type of class. For two unit areas to be classified together to form a region they have to be contiguous. Thus although Köppen identified five areas of Mediterranean climate, they are never considered as being parts of a single region. They are differentiated on the basis of their location. Thus location as a variable is always included in an areal classification or regionalisation.

This contiguity restraint is flexible to a degree. If a river separates two areas forming part of the same class, and would be considered as a part of the same areal class or region were they contiguous, the river is ignored and the areas can be considered as one. This concept can be stretched considerably: for example, islands in the Mediterranean Sea can be considered contiguous with the mainland and therefore within a single Mediterranean region. However, Southern California cannot be conceptually connected with Central Chile or with the large area of South Africa or Southern Europe and therefore each remains a separate Mediterranean-type region. Location is a continuous variable, but we tend to think of the earth as a series of land units.

Occasionally classifications of farms or of units of land are made in such a way that they do not automatically result in regions. We talk loosely of dairy farms or of arable land without extending this to produce dairy regions or arable regions. The processes can therefore be separated: classification in agricultural geography does not necessarily lead to regionalisation; regionalisation is simply one way of providing a classification.

THE PURPOSE AND REQUIREMENTS OF CLASSIFICATION

The purpose of any classification must be clearly understood. Classification is a filter which allows data to be ordered so as to permit further work, the illustration of a theory, or to provide a basis for the testing of an hypothesis. Thus it is essential to have a very clear idea of the particular purpose of a classificatory system before it is adopted. Berry points out that several conditions are assumed to be satisfied before classification proceeds:

> A worthwhile research problem has been posed and formulated, the solution to this problem is facilitated by classification of objects into groups or regions and the data has been collected so there is now an *n* by *m* data matrix.[5]

White, commenting on the failure, in practice, of geographical schemes of land classification for the purposes of river development projects, says:

> In recommending a combined land use and land-use capability survey for the lower Mekong we recognise the danger that it might be so executed as to yield solely a soggy, sterile mass of data. Only by relating it closely to questions of technical and social judgement, and by assuring that the results are in a form usable by the decision makers, could such a survey promise to be genuinely helpful.[6]

Such clarity of purpose has been conspicuously absent in many geographical works. Many of the early attempts to create agricultural regions[7] appear to have been very much ends in themselves.[8] It could be argued that these authors would not accept Bunge's thesis that there is no essential difference between classification and regionalisation and that they were engaged on something more than mere classification, but the purpose of the exercise must be clarified before any attempt is made to regionalise agricultural phenomena. By way of contrast, the purpose of any classificatory system adopted by agricultural economists is usually clear: they classify to assess the number of farms of different types and so make predictions about the future developments of the agricultural sector of the economy.

More precisely, a knowledge of the structure, content and organisation of farming in any area is essential from the point of view of agricultural policy, the material being usually gathered from farm management surveys. A classification can be used to structure the sample of farms needed for this management survey.[9] Four uses for a farm classification have been given by Bennett-Jones:[10]

1. To determine the number of farms of different types so that we can better understand the changes that are constantly occurring in agriculture.

2 To improve the sampling methods used for detailed examination of farming methods and economy.
3 To assist in farm advisory services. It is difficult to judge a farm except against some recognised norm or standard. If a dairy farmer seeks advice in farm management his farm is best compared with other similar dairy farms.
4 As a useful research tool both to narrow the field of study in a manner appropriate to the subject and to allow research to concentrate on a section of the data which is relatively homogeneous.

Equally important is the requirement that the classification chosen be as flexible as possible. It must be capable of being adapted to future developments in the subject. Too rigid a classification structure or too rigid a process of reasoning behind the classification may result in the stereotyping of future research. Although there is little in the geographical literature to lead us to believe that classification of agriculture has been so universally accepted as to threaten this development, there is some evidence that much geographical thought has been ossified in this way in the past. No matter what the nature of the variable being considered, geographers have often felt obliged to study it within regional boundaries.

The difficulties of agricultural classification are increased by the need to impart flexibility into a classification that is based on data collected at a specific time. The problems associated with epitomising agriculture by information collected at a single time of the year have been mentioned in the previous chapter. Changes in the material collected and in the nature of that material may necessitate frequent reclassification.

The result of regional or classificatory thinking may be that the study of aspects of the environment is contained within what may be superficial spatial units and that the interactions of these aspects are confined within these units. Thus purely regional thinking may not only restrict further study but may also lead to seriously erroneous answers. We shall return to this point in a later chapter.

APPROACHES TO CLASSIFICATION

There are two methods of classification: first by the amalgamation of individuals into composite groups, each group being considered as an individual and then amalgamated further; secondly, by dividing the population of individuals into successively smaller groups, each division of the population being chosen according to the presence or absence of some important criterion. The first of these approaches is often inductive. The individuals are searched for similarities. Usually the most similar pair are first grouped together using some measure of the distance between all possible pairs in the population. The most frequently used measure of similarity is the Mahalanobis distance statistic (D^2) developed by Rao[11] and discussed by King.[12] This measure uses the Euclidian distance between two points i and j in n-dimensional space defined by the co-ordinates of the variables being considered. It takes the general form:

$$D_{ij} = (\sum_{n} (S_{in} - S_{jn})^2)^{\frac{1}{2}}$$

where D_{ij} is the summation of all the distances apart of i and j measured in terms of n variables, and S_{in} is the score of the ith individual on the nth variable and S_{jn} is the score of the jth individual on the nth variable.

Using this statistic the most similar pair of individuals can be identified. This pair is then allocated a score on each of the variables included which is determined by taking the mean value of the two individuals now grouped on each of the co-ordinates being used. This group of two then becomes a single individual and the new matrix, one column and row less than the original, is now searched for the next nearest neighbours. This process continues until the individuals have been completely combined into one group. The continuum of generality is then available ranging from the full picture presented by the original individuals to complete generalisation when all individuals form a single group. Some acceptable cut-off point has then to be defined which will be used to create a limited number of classes or groups from the original data matrix.[13] There are other methods of classification by aggregation from below used primarily by plant ecologists but the use of the Mahalanobis distance statistic is the approach most frequently used by geographers.

Classification from above involves dividing the universal set normally into two, on the basis of the most significant variable. For example, in the case of a farm classification it might be appropriate to divide all farms into those which are full-time and those which are part-time on the basis of the amount of total labour inputs required for their operation. Each group can then be subdivided further using other, less important, criteria—for example, rented or owner-occupied categories—until the desired number and size of groups have been arrived at. In this case there would be four groups in all.

In this type of grouping the importance of classification in terms of a body of theory is clear. We need to be able to identify the criteria on which we will divide the population and also the order in which we wish to use these criteria. Different results will be derived by changing either the criteria themselves or the order in which they are used.

The method most often used for this type of classification involves the use of the chi square test to find the attribute that accounts for the largest variance in the population.[14] The population is then subdivided on the basis of this attribute. This type of analysis is most useful where the criteria used for the classification are attributes: for example, farms that are rented or not rented, arable or non-arable, full-time or part-time. Where the criteria are considered as more than attributes, existing as scaled variables, more complicated forms of analysis have to be used. One such form is to divide the population on the basis of the variable which has the highest multiple correlation coefficient with all other variables. In plant ecology this is used where a species is either absent or occurs once, twice, up to n times in a sample quadrat. The quadrats are then divided on the basis of whether or not the plant species with the highest multiple correlation is present or not. Owing to the considerable computation involved in this type of classification it is essential that suitable computer programs are available.[15]

This method of division from above as a form of classification may give rise to theoretical classes, that is a type of farm that could exist but for which we have no example in our sample of farms to fit the necessary requirements. This situation cannot arise from the inductive approach to classification from below which will produce a classification more closely limited by the data being classified.

OBJECTIVITY IN CLASSIFICATION

No classification is objective: one can express the criteria for grouping qualitatively or quantitatively.[16] For example, a farm can be classified as predominantly horticultural or can be classed as a horticultural farm because it has over 75 per cent of its total inputs, or profits or some other measure, relating to, or derived from, horticulture. Neither of these methods of categorising the farm is strictly objective. To set a particular percentage level to define a horticultural farm, even if this level has been defined by rigorous statistical processes to detect significant breaks in the distribution of likely values, is to create subjective classes.[17] Decisions have to be made concerning the criteria which will be used to create the classes. Horticulture as a farming occupation might have been excluded in favour of some more composite variable, for example, arable crops. Chisholm[18] points out that a single farm can be classified according to the percentage of land under different crops or according to size, tenure, access to markets and many other criteria. For each of these it is probable that the classes produced will be fundamentally different. The advantage of statistical tests to define groups is not that the resulting classes can be said to be objective and thereby have some claim to be inviolate, but rather that the processes and the regions they define are replicable. The same results will be produced by different people using the same technique and data. The regions produced will also be comparable with those produced by the same methods, and with the same variables, in different parts of the world.

THE AGRICULTURAL REGION

All forms of agricultural classification use units of land area as individuals which are either combined or divided in the classification process. These units are either the operational unit of agriculture, the farm, or some combination of these into larger and in some ways more convenient areas, usually administrative regions. The choice of unit depends on two factors: first the scale of the variation that it is intended to depict; secondly, the units for which the data are available. We have seen that in many countries agricultural data are collected on a farm basis but are released to the public only after amalgamation into larger administrative units. As a result, any large-scale classification, which cannot be based on sample farm surveys, has to rely on the administrative areas for which data are available to serve as the units which are then classified: for example, parishes or rural districts in the United Kingdom, counties in the United States, and parishes, counties and other combinations of parishes in Sweden.

The size of the areas for which the agricultural data are available controls the mesh size of the 'net' used to capture variation in agricultural variables. No variation smaller than that provided by the agricultural data units will be shown in regions using these data units: parish agricultural data must be a generalisation of the variation between farms within that parish; variations on the individual farm are lost by using a parish mesh for data collection.

Attempts to produce world agricultural regions[19] clearly both use different data—because to collect and process agricultural statistics on the scale of the United States county or the English parish for the whole world would be inconceivable—and produce regions the level of generalisation of which is very different from that of small-scale studies. Between these two extremes lies a whole range of studies on continental or national scales.

Symons[20] uses Whittlesey's classification as a framework and picks on regional- or national-scale examples for many types. Whittlesey's commercial plantation crop tillage is illustrated by a study of plantation agriculture in Malaya while collectivisation in the USSR is examined as an alternative to the plantation. In such studies as this the text tends to be descriptive while the maps must be generalised. The information given about agriculture must be limited by the chosen scale of the investigation.

At the other extreme, in his study of the Isle of Man, only some 57,000 hectares in extent, Birch[21] concludes that there is a need to show the differences in farming patterns within the unit of the parish, of which there are seventeen in his chosen study area. The field is used as the smallest unit for the collection of land-use data and a sample of farms provides other farm data. Combined, these provide information for the delimitation of farming-type regions.

AGRICULTURAL CLASSIFICATION

It is possible to identify three basic approaches to classification in agricultural geography depending on the criteria to be classified and therefore the type of regions that will be produced.

1. Land classification, where the land itself is classified and regions are created based on its physical properties or capabilities.

2 Land-use classification, where the use to which the land is put is used as the basis for regional division.
3 Type of farming classification, where the actual farming operation is the basis for the classification.

LAND CLASSIFICATION

No claimants to the use of land present their case more coherently than do agriculturalists. The land classification map is the most common way of presenting their case to the planners. Land classification can take many forms, perhaps the most extensively used being that adopted by the Commonwealth Scientific and Industrial Organisation (CSIRO) for the physical classification of land in Australia and adopted widely throughout the world.[22] The land is broken into a number of units within which there is a recurring pattern of topography, soils and vegetation. These units are then used to illustrate agricultural capability and use.[23]

The first major attempt in modern times to classify the land of Britain, in terms of its physical properties in relation to agriculture, was produced by Stamp[24] whose classification stemmed from the Report of the Committee on Land Utilisation in Rural Areas known as the Scott report.[25] During the preparation of this report it became clear that there was no satisfactory basis on which to make decisions concerning the appropriation of agricultural land for other purposes. There was no scale against which any plot of land could be measured to assess the strength of the case for retaining it in agricultural use. There was, of course, considerable feeling during and after World War II for the preservation of agricultural land to encourage self-sufficiency in food production within the United Kingdom. It was in response to this need that the Scott Committee asked the Land Utilisation Survey to prepare such a classification of land. The basis of this survey was to be the land's inherent fertility and productivity. Using the material collected by the land-use survey and published in the various county reports, a threefold classification of land was drawn up. These three major classes—good, medium and poor—had various subdivisions depending on the type of agriculture the land encouraged (Table 9).

In the early 1940s the Soil Survey of England and Wales criticised this classification as over-emphasising the importance of existing use rather than potential use as the original statement of

TABLE 9 Land classification, England and Wales, 1948

Major category	*Sub-category*	*Percentage of total area of Great Britain*
Good		37·9
	1. First class	4·1
	2. Good general purpose farmland	
	a. Suitable for ploughing	15·2
	g. Suitable for grass	5·0
	3. First class land, restricted use, unsuitable for ploughing	2·2
	4. Good but heavy land	11·4
Medium		24·6
	5. Medium light land	
	a. Suitable for ploughing	4·4
	g. Unsuitable for ploughing	0·4
	6. Medium general purpose farmland	19·8
Poor		35·2
	7. Poor heavy land	1·6
	8. Poor mountain and moorland	31.7
	9. Poor light land	1·5
	10. Poorest land	0·4
Residue		
	11. Built-up area	2·3

(Based on Stamp, L. D., *The land of Britain: its use and misuse*, 1948)

intent had implied. In view of the background of the Land Utilisation Survey (see Chapter 3) this criticism is not surprising. However, the Soil Survey predicted that about 50–60 years would be needed to achieve the detailed knowledge of soils needed for a potential land-use classification.

While there is little doubt that soil is an important physical variable affecting patterns of agriculture, it is not by any means the only such factor.[26] For a suitable physical classification details are also needed concerning relief, climate, and hydrology. While the last three factors have considerable influence on soils and, therefore, are reflected in soils, these three factors also affect agriculture in a direct manner such that many different forms of agricultural practice can exist on the same soil type.

In response to a growing need for an up-to-date map of the United Kingdom showing land capability the Ministry of Agriculture has produced a series of maps at one inch to a mile by selecting certain critical levels in respect of soil (depth, structure, chemical composition and permeability) and other physical criteria (slope, precipitation, drainage, temperature, susceptibility to frost, availability of ground water). These are used to assess the limitations of the physical environment, and the sum of physical limitations in a particular area gives an indication of the land capability. These classes are then graded from very suitable to highly unsuitable for agriculture and the details are mapped.

A similar, but more detailed, sevenfold classification of land has been used in a report for multiple land-use planning in Ontario 'for the purpose of land administration and for land-use planning in general'.[27] Soil forms the basis of the classification, and land is classified according to the costs of developing the land for commercial agriculture. As it is typical of this form of classification it is reproduced in detail.

1. Agricultural use capability: Class A

These are lands with very low development costs and low to moderately low maintenance costs. Representative soils of Class A are clay and loam of good structure and comparatively free from compact layers which limit root and water penetration. The natural drainage is good. There are few restrictions as to stoniness, erodibility, etc. Class A lands are not seriously broken by rock outcrops, swamps, muskegs, steep slopes and eroded areas. In brief, Class A lands are those which can be brought up to satisfactory agricultural production with the lowest costs and the simplest methods of any land within the region, though this low cost is not necessarily economical.

Specifically, Class A lands require some, if not all, of the following practices before they will produce general farm crops on a level satisfactory for the region:

(i) minor improvements in drainage;
(ii) moderately low improvements in soil fertility;
(iii) simple erosion control methods.

2. Agricultural use capability: Class B

These are lands with low developmental costs and low to moderately low maintenance costs. The greater developmental costs compared with Class A are due to one or more of the following:

(i) greater drainage requirements;
(ii) an increase in stoniness;

(iii) coarser and poorer soil;
(iv) steeper slopes.

The following types of land are representative of Class B:

(i) clay and loam soils with imperfect natural drainage comparatively free from stones and compact layers and comparatively unbroken by rock outcrops, steep slopes, muskegs, etc;
(ii) clay and loam soils with good natural drainage which are somewhat difficult to develop because one of the following features interferes to a moderately low degree: stoniness; compact layers; rock outcrops; steep slopes or other erosion features; muskegs or other areas of difficult drainage.

Specifically, Class B lands require some, if not all, of the following practices before they will produce general farm crops on a level satisfactory for the region:

(i) moderately low improvement in drainage;
(ii) moderately low to complex improvement in soil aeration and fertility;
(iii) simple to moderately complex erosion control practices;
(iv) removal of few stones as required.

3. Agricultural use capability: Class C

These are lands requiring moderately low expenditures for development and low to moderate expenditures for maintenance. For example, the fine-textured soils of the Cochrane Clay Belt with 6–18 inches of peat are included in this class.

The following types of land are representative of agricultural use capability Class C:

(i) clay and loam soils with imperfect natural drainage which are moderately difficult to develop because one of the following features interferes to a moderate degree: stoniness; compact layers; rock outcrops; steep slopes or other erosion features; muskegs or other areas of difficult drainage.

Specifically, Class C lands require some, if not all, of the following practices before they will produce general farm crops on a satisfactory level:

(i) moderately low drainage costs;
(ii) moderately low methods in the improvement of soil structure, aeration and fertility in general;

(iii) simple erosion control practices where necessary;
(iv) low cost of removing stones.

4. Agricultural use capability: Class D

These are lands with moderate costs in both development and maintenance. The clay soils of the Cochrane Clay Belt with 18–36 inches of peat are in this class. Other commonly occurring examples are:

(i) clay and loam soils which are difficult to develop because one of the following features interferes to a moderately high degree: stoniness; compact layers; rock outcrops; steep slopes or other erosion features;
(ii) imperfectly to poorly drained clays with a shallow covering of sand;
(iii) clay soils comparatively shallow over bedrock.

Class D lands may be so rated also because of the distribution pattern of good and poor soils, for example, clay and loam areas broken by rock, sand spots or poorly drained areas.

Specifically, Class D lands require some of the following practices before they can produce general farm crops on a satisfactory level:

(i) complex drainage operations including, in many cases, the removal of considerable depth of peat;
(ii) complex soil improvement practices often of a physical nature such as breaking up compact subsoil;
(iii) complex improvement in soil aeration and fertility;
(iv) low to moderate erosion control practices;
(v) moderate costs of removing stones.

5. Agricultural use capability: Class E

These are lands with moderately high costs of development and maintenance. In some cases it is the maintenance cost rather than development cost which places land in this category. A common example is a coarse-textured soil. The cost of maintenance of these soils continues to be higher, although the cost of development may not exceed that of Class D. Included in Class E also are the organic soils deeper than three feet but reclaimable with moderately high costs.

The following types of land are included in Class E:

(i) coarse-textured soils with a smooth to gently rolling topography, with imperfect to excessive drainage, and comparatively free from stones and unbroken by rock outcrops, steep slopes, eroded areas and muskegs, etc;
(ii) fine and intermediate textured soils, moderately stony and comparatively shallow over bedrock, having a smooth to gently rolling topography.

Specifically, Class E lands require some, if not all, of the following practices before they will produce farm crops on a level satisfactory for the region:

(i) heavy applications of mineral and organic fertilisers and of mineral and organic material (such as lime and peat);
(ii) removal of stones in the fine textured soils;
(iii) moderately complex erosion control methods.

6. *Agricultural use capability: Class F*

These are lands with high developmental and maintenance costs. They are the very shallow, the very stony, very steep and severely eroded lands.

Lands which are badly broken by bare bedrock, steep ridges or bogs are included in this class, even though the remainder of the land could be rated Class E and higher.

7. *Agricultural use capability: Class G*

These are lands with bare bedrock, open bogs, and extremely broken lands on which the costs of development and maintenance are very high. These cannot be reclaimed except by very expensive engineering projects.

A more restricted classification has been devised by the United States Soil Conservation Service. It is concerned mainly with the land's susceptibility to soil erosion, taking into account such parameters as slope and aridity, but Symons[28] concludes that it has very limited value beyond the narrow consideration of soil erosion as it pays little heed to more general features of productivity.

The complexities of producing maps which depict the physical capability of land to produce all agricultural crops can be overcome by restricting attention to single crops or limited combinations of crops. This approach is recommended by Hilton[29] and developed by O'Connor[30] for horticultural crops.

Any map showing physical capability of land will lead to many anomalies when the map is compared with the real world. Good and bad quality farming is not restricted to good and bad quality land respectively. The type of farming structure is also important in any classification of agricultural land. Hilton[31] suggests a method based on that used by the Cornell University land surveyors[32] and common throughout the eastern seaboard of the United States. Essentially the land is classified according to the economic production data modified in terms of critical physical phenomena. The classes are then modified further in terms of economic efficiency. This method of land classification can be traced back to the work of Hudson and his colleagues at the Tennessee Valley Authority in the mid-1930s.[33] In this study each homogeneous unit of land of not less than 200 acres was allocated to one of five classes on the basis of the agricultural quality of the present physical conditions of the land. Each unit was also placed into one of five classes on the basis of the quality of the present agricultural use of the land taking into account such variables as the size of fields, cover of weeds, scrub and bush, proportion of idle land and the quality of the farmsteads and equipment. Finally each unit was allocated to one of five classes on the basis of the economic status of the farmers and the physical conditions of the land. The final classification rested on the extent to which agricultural problems were present, whether physical, social or economic. In the United Kingdom the only measure available to Hilton as an indication of efficiency was farm size, whereas the Cornell land surveyors were able to use certain visual features of the farm. The Cornell surveyors also subjected the classes that they determined to rigorous testing and modified the resulting boundaries if necessary. The United Kingdom examples could not be so checked because of the dearth of farm management statistics; none the less the resulting maps proved acceptable in all but a few details when examined by local officials of the Ministry of Agriculture and by local farmers.

The main advantage of a purely physical classification of land lies in its more or less permanent nature[34] which can serve as a comparison with the more rapidly changing social and economic characteristics. On the other hand it is clear that agriculture is much more than the sum of the physical properties of land. It is a combination of physical capabilities and man's reaction to them and his ability to overcome them. Although the physical properties of land may change slowly they do not remain unaffected by man's activities. The very physical factors which Hilton regards as more or less permanent may themselves be the result of man's activity. Clearly man and the land are intimately connected in this respect. In addition, there is little doubt that land is decreasing in importance as a control over the type of farming. This is especially true in terms of the fertility controls which it imparts. Man is now able to so control and alter the physical properties of his environment that land assumes the position of one of many *economic* controls on the pattern of agriculture.

In concluding this section it should be noted that these exercises in land classification have been no more than the grouping of individual parcels of land with no contiguity restraints. The classes were not formalised into regions but simply mapped as they stood with no more attempt at generalisation than was required by the scale of the mapping exercise. As a result the land classification maps produced by the British land-use survey and by the United Kingdom Ministry of Agriculture show complex intermixtures of land classes within very limited areas.

LAND-USE CLASSIFICATION

It has often been thought more appropriate by geographers to classify the use to which the land is put rather than its physical characteristics. We have already considered the land-use classifications which were essential prerequisites for land-use survey work which could then be used for a land resource inventory. This section deals with those classifications of land use which are derived from the data rather than those which precede its collection.

The greatest exponent of this form of land-use classification was J. C. Weaver.[35] Weaver can be distinguished from his contemporaries in that he realised that the most important feature of agricultural distributions was that crops are grown in association with each other, not in isolation. Further, he devised a statistical method of representing these associations. Earlier, the consideration of single crops had led to the delineation of such allegedly characteristic single-crop regions as the corn belt, the cotton belt and the spring wheat belt in the United States. R. O. Buchanan[36] echoes the views of Weaver:

> The Australian sheep areas, the New Zealand specialist dairy farming regions, the coffee belt of Brazil, the rubber areas of Malaya, may well exemplify extreme concentration on a single product, but the corn belt of the United States, still more the mixed farming area of East Anglia, have no such limitations.

Weaver recognised that such unities did not exist but that rather, throughout the corn belt to which he devoted much of his energies, there was considerable variation. Corn was *not* the dominant crop throughout the belt and it was grown with a variety of other crops. What appeared to Weaver to be critical was the association of crops found together. Weaver's purpose was to delimit not single-crop regions but crop-combination regions.

The procedure is very simple. It involves the selection of crops which are to be included in the analysis and necessary amalgamations of these into larger groups of crops if this is convenient. The selected crops in each area are then represented as a percentage of the total cropped land in that area. This preliminary selection and grouping illustrates the point that no classification system, however mathematically sophisticated, can be totally objective in character. Weaver,[37] in his original study, was forced by inadequate data to omit pasture from his analysis. Both permanent grass and temporary leys should normally be considered as important as any other crop used in a rotation system. Also, certain types of crops could never hope to be represented if considered on their own, but if grouped into larger units can become a significant component of the total agricultural acreage in the area being considered. For example, carrots would not normally form a significant component of total agricultural acreage in the United Kingdom. If, however, they are combined with all other vegetables and horticultural crops then this combination becomes a formidable unit in much of eastern England. The results of any classification, particularly if based on percentage data, will be substantially altered by changes in the make-up of the individual crops being classified. It is essential, therefore, when comparing studies using Weaver's system of classification, to make sure that the basis of the classification has been the same in all cases.

The crops or groups of crops are arranged in order according to the percentage of the total agricultural area which they cover and the problem is to select only those crops which are sufficiently important in the local agricultural system to determine the cropping pattern adopted.

Weaver then compared the actual range of percentage values with a series of model situations. In a system of monoculture all the land would, by definition, be devoted to a single crop. If there was an ideal two-crop system, 50 per cent of the land would be given over to each crop. We can continue with these model situations until there are as many crops in the model as we have recorded data for. This is illustrated in Table 10 and Figure 19 taking the parish of Acle in Norfolk in 1968 as the example.

The next stage is to find which of these ideal models fits the actual system most closely. This is done by a simple measure where the difference between the percentage of land under each crop, and that predicted for it by the model, is squared and added to the differences calculated in the same way for all the other crops being considered. Thus:

$$W = \sum_{i=1}^{n} (p_{oi} - p_{ei})^2$$

where p_{oi} is the observed percentage of land devoted to crop i and p_{ei} is the expected percentage by the model. W provides a measure of the goodness of fit of the model. Weaver's original method was amended by Thomas[38] to include in this W statistic all the crops while Weaver considered only those crops which were actually included in the model being used. In this case Weaver would reach a D^2 value of 864·4 for Acle using the two-crop model where n equals two whereas the modified method with n equal to nine gives a figure of 1,083·8. The lowest of the W values gives us the model to which the real cropping pattern approximates most closely. We are thus able to identify Acle parish as a two-crop system consisting of permanent grass and barley grown in combination.

The crop combination regions are then mapped using shading to depict the dominant crop and letters overprinted to illustrate the subsidiary ones within the system in each area. The method is relatively crude as the remaining difference between reality and the closest model may be fairly large. Weaver showed the degree of fit of the model in each case by drawing isopleths of

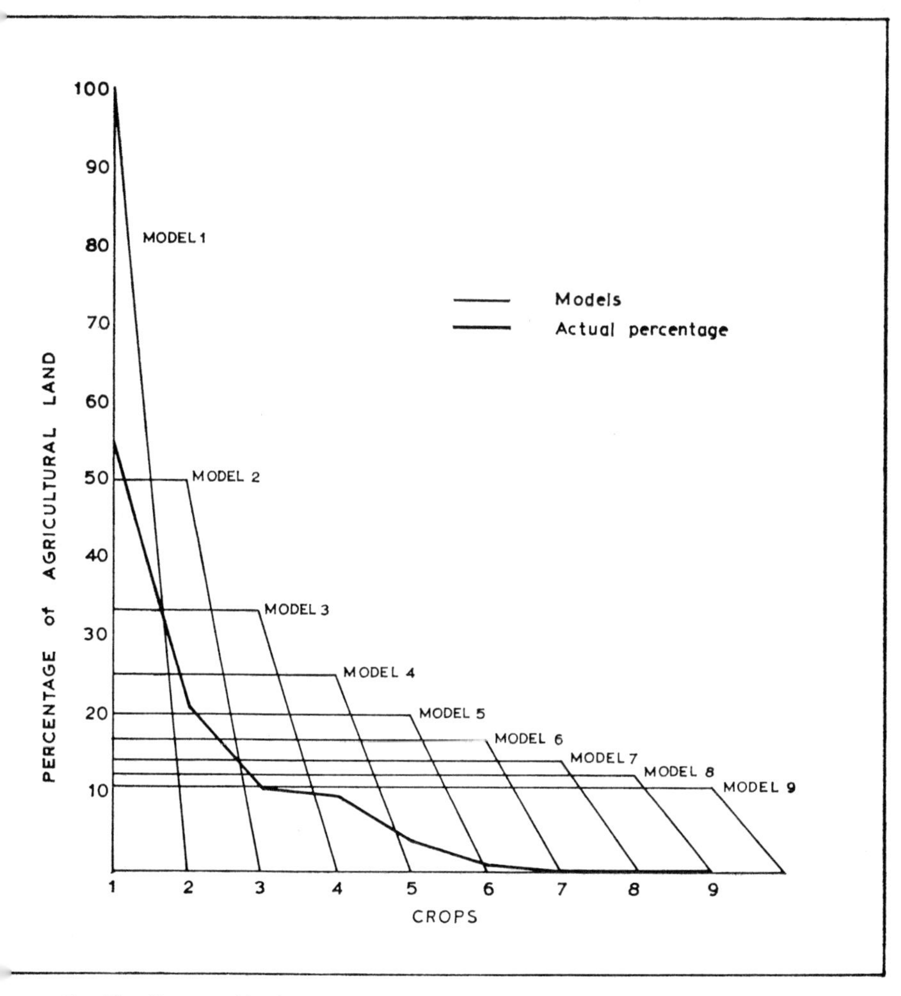

Fig 19. Crop combination models for the parish of Acle in 1968. The actual percentages of the land of the parish under the leading ten crops are indicated by the heavy line. The various model situations, assuming only one crop, two crops, etc, are indicated by the nine thin lines. The two-crop model is found to show the least deviation from the actual pattern in this case.

TABLE 10 Crop combinations, Acle parish, Norfolk, 1968

Crop acreage	%	*1 crop model*	D^2	*2 crop model*	D^2	*3 crop model*	D^2	*4 crop model*	D^2	*5 crop model*	D^2	*6 crop model*	D^2	*7 crop model*	D^2	*8 crop model*	D^2	*9 crop model*	D^2
Grass 1827	54·2	100	2,097·6	50	17·6	33·3	436·8	25	852·6	20	1,169·6	16·7	1,408·7	14·3	1,593·1	12·5	1,738·9	11·1	1,856·6
Barley 706	20·9	0	436·8	50	846·8	33·3	153·8	25	16·8	20	0·8	16·7	17·9	14·3	43·7	12·5	70·6	11·1	95·8
Wheat 349·25	10·3	0	106·1	0	106·1	33·3	529·0	25	216·1	20	94·1	16·7	40·5	14·3	15·9	12·5	4·8	11·1	0·66
Sugar beet 333·25	9·9	0	98·0	0	98·0	0	98·0	25	228·0	20	102·0	16·7	45·8	14·3	19·2	12·5	6·8	11·1	1·5
Vegetables 131·0	3·9	0	15·2	0	15·2	0	15·2	0	15·2	20	259·2	16·7	163·0	14·3	107·9	12·5	74·0	11·1	52·0
Small fruit orchards 10·0	0·3	0	0·1	0	0·1	0	0·1	0	0·1	0	0·1	16·7	267·9	14·3	195·6	12·5	148·8	11·1	116·9
Potatoes 7·0	0·2	0	0	0	0	0	0	0	0	0	0	0	0	14·3	178·4	12·5	151·3	11·1	119·1
Stock feeding 5·75	0·2	0	0	0	0	0	0	0	0	0	0	0	0	0	0	12·5	151·3	11·1	119·1
Others 3·25	0·1	0	0	0	0	0	0	0	0	0	0	0	0	0	0		0	11·1	121·2
Total 3,372·5	100		D^2 2,753·8		D^2 1,083·8		D^2 1,232·9		D^2 1,328·8		D^2 1,625·8		D^2 1,943·8		D^2 2,173·8		D^2 2,346·5		D^2 2,482·9

D^2 for Weaver model

1 = 2,079·6
2 = 864·4
3 = 1,119·6

(Based on data published by the Ministry of Agriculture, 1968)

the deviation from the ideal model divided by the number of crops in the system for each county.

Using agricultural data by parish for the county of Norfolk the pattern of crop combinations can be seen in Figure 20. The leading crops in terms of their labour requirements are vegetables, sugar beet and potatoes while in a limited number of parishes wheat becomes dominant. Most of central Norfolk has a two- three- or four-crop system based on sugar beet as the most important crop and usually with barley as the second most important. In eastern Norfolk vegetables become the leading crop in predominantly three-crop systems with sugar beet, potatoes and barley. In western Norfolk that portion of the Fens within the county has vegetables as the leading crop with potatoes often appearing as of secondary importance. To the east of this area is a band of parishes where potatoes are usually the major crop in what is often a very diverse agriculture; five-crop systems are not uncommon. This area is a type of transition zone where many of the parishes include a number of soil environments based on boulderclay, chalk, greensand, gault clay and fenland peat.

This type of map shows not only the distribution of major crops but also the degree of specialisation or diversity in the agriculture of an area. As the method of measurement is based upon labour demands rather than land area used for each crop, extensive land-use types such as grass reach significant levels only in limited areas, in the case of Norfolk only in one parish in Broadland in the east of the county.

A problem with this method is the representation of speciality crops which may never achieve a prominence sufficient to be included within the combination system, but which are sufficiently distinctive to make their mark on the local agricultural patterns. Weaver limited his study to a consideration of nine field crops but overprinted on his maps speciality crops where they achieved considerable importance locally. In these studies the usual marginal problems are represented by the case of rye. Weaver was comparing patterns of field crops in the corn belt area for the years 1919, 1929, 1939 and 1949.[39] In 1949, rye was of no importance and was considered as a speciality crop in some small areas. But in 1919 rye was of considerable importance over large areas of the mid-west if measured in terms of the total acreage that it occupied. Had it been included as a major crop in the original analysis for 1949 there would almost certainly have been a

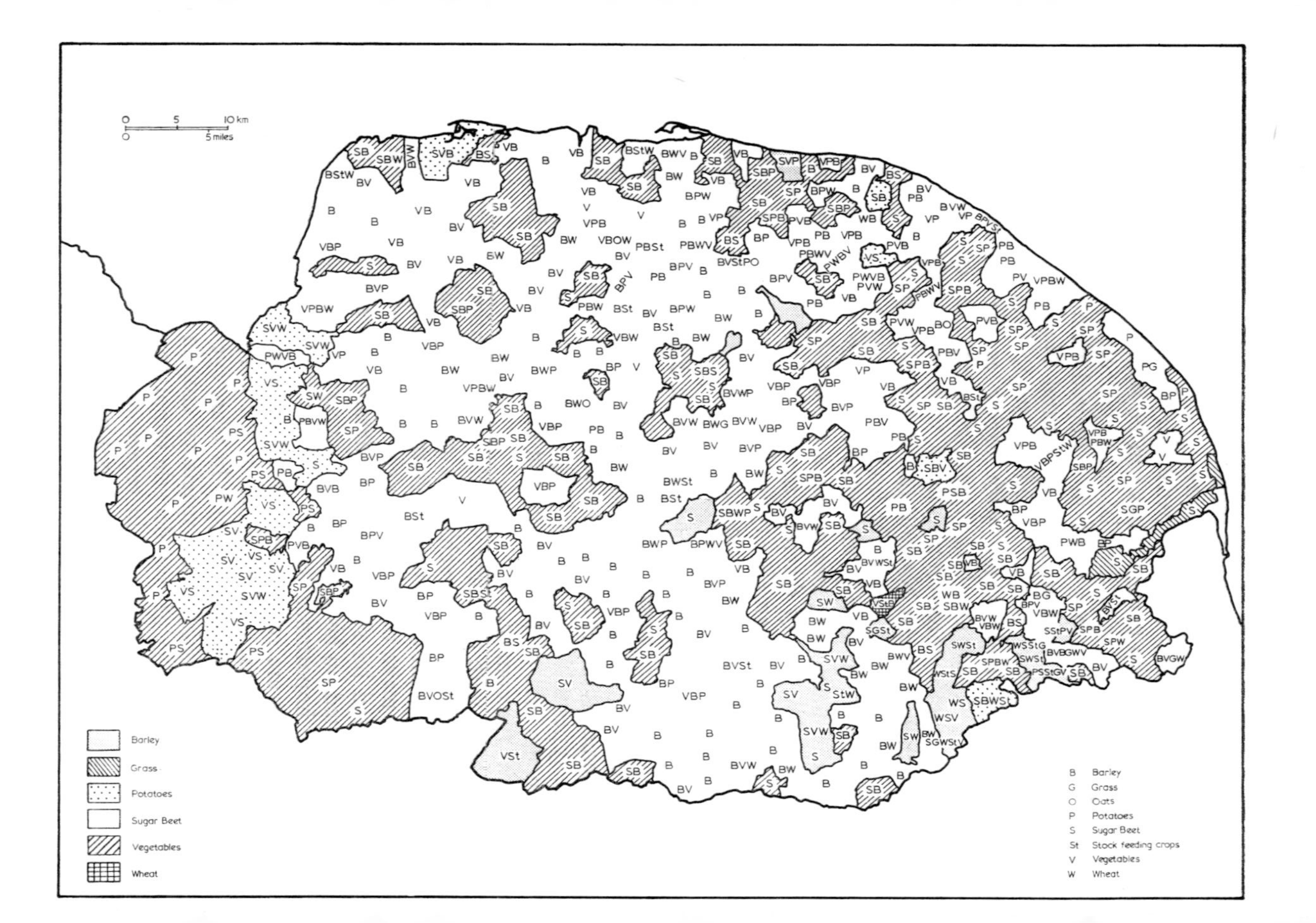
0 5 10 km
0 5 miles
Barley
Grass
Potatoes
Sugar Beet
Vegetables
Wheat
B Barley
G Grass
O Oats
P Potatoes
S Sugar Beet
St Stock feeding crops
V Vegetables
W Wheat

greater diversity of crop combinations, especially in the earlier years.

Weaver's method has been taken up and refined by others: for example, Scott[40] used it to assist in the definition of agricultural regions in Tasmania; Coppock[41] has used it extensively in England and Wales. In particular, the method is not limited to crop data but can be used to classify areas according to any set of data which can be represented by percentages. It could, for example, be used to classify areas according to the employment structure found within them. Within agriculture it has been used by Weaver and others to produce classifications of areas based on agricultural livestock. All livestock are reduced to standard livestock units based on feed requirements and then each type is expressed as a percentage of the total livestock units for that area. The livestock conversion units are shown in Table 11, and Table 12 illustrates the method applied to Acle parish. Acle parish is now identified as an area with a two-crop system and a single livestock system of cattle superimposed upon it.

TABLE 11 Livestock conversion units

Type of livestock	*No. of livestock feed units**
Horses	1
Cows, bulls and other cattle 2 yrs old and over	1
Other cattle 1 yr old and under 2	2/3
Other cattle under 1 yr old	1/3
Breeding ewes (including shearling ewes)	1/5
Rams	1/10
Other sheep	1/15
Sows	1/2
Boars	1/4
Other pigs	1/7
Poultry 6 months old and over	1/50
Poultry under 6 months old	1/200

*These figures have been in use in England and Wales; in countries where different stock rearing systems apply these figures will not be applicable.
(After Coppock, J. T., *Agricultural atlas of England and Wales*, 1964)

It must be remembered that Weaver[42] admits a failure to superimpose successfully the results of his livestock combination regionalisation upon the crop combination regions and thereby produce a map of farming regions. Although the production of both crops and livestock on the same farm, and certainly in the same region, is common, there appeared to be no statistical method for Weaver to combine both in a single index.

Thus Chisholm concludes that:

> despite his Herculean labours, we are no nearer an understanding of the regional relationships between crops and livestock, one important aspect of which is the amount and nature of the arable crops fed to the livestock. We do not know whether it is corn or hogs that provide the main source of income in the area he studies.[43]

TABLE 12 Livestock combinations, Acle parish, Norfolk, 1968

Livestock	*No.*	*Livestock units*	*%*	*1-stock model*	*D^2*	*2-stock model*	*D^2*
Cattle	514	514					
	291	194					
	400	133·3					
Total cattle	1,205	841·3	88·9	100	123·2	50	1,513·2
Pigs	58	29					
	4	1					
	241	34·4					
Total pigs	303	64·4	6·8	0	46·2	50	1,866·2
Poultry	465	9·3					
	6,246	31·2					
Total poultry	6,711	40·5	4·3	0	18·5	0	18·5
Sheep	0	0	0	0	0	0	0
Total livestock	8,219	946·2	100		187·9		3,397·9

(Based on data published by the Ministry of Agriculture, 1968)

TYPE OF FARMING REGIONS

Chisholm[44] lays down particularly stringent requirements for a type of farming region where the individual units classified must be the farms themselves and where the classification must include all variables concerned with the production and management of the farm as well as information on yields, crops and livestock. There is no doubt a danger in equating simple land use with type of farming. This type of confusion is perpetuated by such publications as the *World Atlas of Agriculture*[45] which presents maps showing the distribution of categories of land use: arable; meadows and permanent grass; fruit trees, vineyards, bushes and orchard land; market gardens, and nursery gardens; woods and forests; rough grazing; non-agricultural land. This *Atlas* is a portrayal of land use, though there are a number of illustrated monographs on the agricultural economy of the various countries to go alongside these maps.

Chisholm concludes that the comments of Hartshorne and Dicken[46] concerning the apparent control of the physical environment and the unsignificance of market location for the location of grain production in Europe were erroneous because they took no account of the methods of production and yields while claiming to consider types of farming region: Hartshorne and Dicken were confusing land use with type of farming. On the other hand we have seen that no classification exists which attempts to include information on both livestock and cropping patterns and we have noted Weaver's comments on the desirability, but apparent impossibility, of doing this. It has become generally understood that classifications, at least based on the twin criteria of crops and livestock, no matter what the unit of measurement or what scale of area is used, are called type of farming regions.[47] Chisholm's more stringent requirements then become a more advanced and rigorous classification of the same general type.

The most widely accepted classification of agriculture on a worldwide scale has been that adopted by Whittlesey.[48] The most important assumption of this work was that agricultural regions should be defined on the basis of agricultural phenomena, a feeling echoed by Buchanan.[49] Whittlesey identified five features of agriculture which were important for the definition of world agricultural regions:

1. crop and livestock associations;
2. the intensity of land use;
3. the processing and disposal of farm produce;
4. the methods and degree of farm mechanisation;
5. the types and associations of buildings and other structures associated with agriculture.

From these criteria Whittlesey derived thirteen types of world agriculture:

1. nomadic herding;
2. livestock ranching;
3. shifting cultivation;
4. rudimentary sedentary tillage;
5. intensive subsistence tillage with rice dominant;
6. intensive subsistence tillage without paddy rice;

7. commercial plantation crop tillage;
8. Mediterranean agriculture;
9. commercial grain farming;
10. commercial livestock and crop farming;
11. subsistence crop and stock farming;
12. commercial dairy farming;
13. specialised horticulture.

It is not possible to differentiate most of these classes statistically. For example, it is impossible to place shifting cultivation on a point of a quantitative scale including paddy cultivation and dry farming. While these may be three different intensities of land use the classes remain descriptive. Despite its apparent lack of rigour, Whittlesey's classification became generally adopted with relatively minor modification. Symons[50] has used this framework and provides a description of each of Whittlesey's classes before considering a limited number in considerable detail.

Helburn[51] has argued for a radical rethinking of classification of world agriculture so that it can be based on variables that are quantitatively measurable. Three of Whittlesey's original criteria survive: the degree of dependence on different crops and livestock, the intensity of land use, and the proportion of the production sold commercially. To these he adds eight further variables which either can now be quantified, or might in the future with improved statistical collection be quantifiable:

1. degree of specialisation;
2. labour and capital ratios to land and to each other;
3. sedentary as against migratory habits;
4. scale of operation;
5. land tenure systems;
6. level of living achieved;
7. value of the land;
8. value or volume of production.

If there were to be three arbitrary classes to each type of variable it would be possible to find at least 59,000 possible combinations or types of agriculture! Clearly it is necessary to isolate from the above list the important variables for a primary classification. Helburn chooses to emphasise the relation of crops and livestock, labour and capital to each other and to land, and the degree of commercialisation. The classes then derived can be subdivided according to the degree of specialisation, sedentary against migratory habits, or any other variables which appear to be significant. A classification of this type depends on widespread data collection on a uniform basis and though Helburn is optimistic that the demand for this data will ultimately lead to its supply this has not yet been achieved.

Moving from the world scale to regional studies, pioneering work was published by O. E. Baker on the agricultural regions of the United States for several years from 1926 onwards. Stemming from this work came a series of studies covering all other continents along similar lines to those adopted by Baker.[52] The basis of Baker's studies was the statistical unit of the American county. These units are sufficiently small and the statistics sufficiently detailed for definition of regional boundaries on the scale of a country the size of the United States. The object was to provide a degree of precision to the boundaries of agricultural regions of the United States, many of which, particularly the cotton and the corn belts, were in common everyday usage. Certain of these boundaries could be defined as critical climatic limits beyond which the particular crop was not grown, for example, the northern and the western boundaries to the cotton belt: in such cases there was little dispute. In other cases Baker's methods appear less precise and criticism has been made of his corn belt boundaries by R. O. Buchanan.[53] Despite the detailed criticism since levelled at Baker and his contemporaries, there is little doubt that the enormous body of literature they produced on agricultural regions stimulated considerable thought about the distribution patterns and possible regional divisions of agriculture. Their work makes up a substantial part of the early literature in economic geography.

Unfortunately, more typical of attempts to regionalise agriculture on a national scale are those which use non-agricultural criteria to provide the basic regions within which the work is reduced to a description of agricultural practices. In the studies of agriculture of the United Kingdom edited by J. P. Maxton,[54] the country is divided into fourteen areas which are amalgamations of counties such as 'the Eastern Counties, consisting of Norfolk, Suffolk, Essex, Hertfordshire, Bedfordshire, Huntingdon, Cambridgeshire and the Holland and Kesteven divisions of Lincolnshire'. Each of these fourteen areas is systematically described. This type of approach is followed in numerous geographical texts when dealing with agriculture, an approach that Whittlesey tried to counter. The essential feature of any classification of areas according to agriculture, or type of farming, must be that the bases of the classification are agricultural and that they attempt to include as many aspects of agriculture as possible, at the very least a consideration of crops and livestock together.

To permit any classification to include a statistical analysis based on both crops and livestock some conversion factors must be found so that the acreage of different crops can be compared with the number of different types of livestock. As we have seen, crops can be standardised by calculating each as a percentage of the total agricultural land. Similarly livestock can be standardised by converting all to numbers of livestock units. It is now important to combine these two in some way. The problem is made more acute by the different intensities under which different crops are grown. To compare an acre of wheat with an acre of sugar beet or an acre of peas grown for canning is to obscure the financial yield per unit area of the crops.

There are several different methods of standardising the different agricultural variables so that they can be combined into the same regionalisation procedure. The choice between them depends largely on the availability of suitable data and partly on the objectives of the classification in the first place. Three methods have been suggested by Adeemy[55] and by Coppock:[56]

1. Standard man-day units

This is the most common method of standardisation adopted by geographers in the United Kingdom and by the United Kingdom Ministry of Agriculture. It has also been used as the basis for allocating aid to small farmers.[57] With this method all types of production on the farm are reduced to their standard labour requirements. The obvious problem with the use of such a measure is that there is no such thing as a standard farm. The method does not allow for the gradations in efficiency of different farmers nor for such changes as are brought about through economies of scale. Also, labour inputs are not the only inputs on the farm and changes in capital investment and fertiliser application may well have substantial effects on the labour requirements of particular farms. The standard man-day requirements of most types of United Kingdom agricultural production are shown in Table 13. These are the most up-to-date currently available in the

TABLE 13 Standard labour requirements, England and Wales, 1971

Crops	*Standard man-days (per acre)*
Wheat	2
Barley	2
Oats	3
Mixed corn	3
Rye—for threshing	2·5
Potatoes—first early	15
—main crop second early	15
Beans for stock feeding	3
Turnips, swedes and fodder beet for stock feeding	9
Mangolds	11
Rape or cole for stock feeding	1
Kale for stock feeding	1·5
Cabbage, savoys and kohl rabi for stock feeding	5
Mustard for seed, fodder or ploughing-in	3
Other crops for stock feeding excluding lucerne and grasses	3
Sugar beet	10
Hops	70
Orchards grown commercially	23
Orchards not grown commercially	1
Small fruit:	
Strawberries	70
Raspberries	80
Currants, black	40
Gooseberries	40
Other small fruit	60
Vines (grapes)	50
Vegetables for human consumption:	
All brassicas	20
Carrots, early	60
Carrots, main crop	10
Parsnips	25
Turnips and swedes	20
Beetroot (red beet)	30
Onions, salad	100
Onions, harvesting dry	25
Beans, broad	30
Beans, runner: bush	30
Beans, runner: climbing	80
Beans, french	25
Peas, green for market	30
Peas, all other for processing	3
Celery	40
Lettuce, not under glass	30
Other vegetables and mixed areas	50

TABLE 13 continued

Crops	*Standard man-days (per acre)*
Chicory	11
Sage	25
Watercress (per 100 sq ft)	1
Hardy nursery stock	50
Bulbs	100
Other flowers not under glass	250
Crops under glass or sheds	1,300
Other crops not for feeding to farm livestock	3
Bare fallow	0·5
Lucerne	1·25
Clover, sainfoin and temporary grasses	0·75
Permanent grass	0·5

Livestock	*Standard man-days (per head)*
Dairy cows in milk or in calf and heifers in milk	10
Dairy heifers in calf	3·5
Beef cows in milk or in calf and heifers in milk	3
Bulls being used for service	6
Bulls (at A.I. centres)	20
All other cattle	2·5
Goats—milch goats	5
—other goats	2
Ewes	0·7
Rams	0·7
Wethers and other sheep 1 year old and over	0·2
Breeding sows and gilts	4
Boars	4
Other pigs over 2 months old	1
Store pigs kept for some weeks after weaning	0·1
Poultry—Hens and pullets (other than growing pullets), geese and turkeys	0·1
—Fowls for breeding	0·15
—Ducks	0·2
—Broilers, growing pullets and other table fowl	0·05
Day-old chicks	0·003
Mink (female)	1
Rabbits (meat or fur) breeding doe	1
Deer (kept enclosed, fed and culled for meat)	0·5
Bees, per colony	1

NB An addition of 15 per cent should be made for essential maintenance and other indirect labour.

(After data supplied by the Ministry of Agriculture)

United Kingdom but they are frequently changed as farming practices alter. These conversion units are based on knowledge collected by local agricultural advisory officers from their experience in the field and an average figure is then arrived at for the whole country. They are compiled for England and Wales alone and should not be used in other countries where the labour situation in farming may be very different. Crops are no longer ordered according to the amount of land they occupy on the farm but rather in terms of the amount of labour they require. This leads to some drastic changes in the relative importance of different crops. Labour-intensive crops such as sugar beet and more especially vegetables assume considerably increased importance, while labour-extensive crops such as cereals decrease in importance. In the case of livestock the change from ordering by feed requirements to ordering by labour requirements naturally boosts the importance of dairy cattle at the expense of beef. Acle parish is again used as an example of the changes that these conversions make from the figures derived using total crop acreages and livestock units (Table 14).

2. Gross margin data

In this case the gross output is calculated, minus all variable costs of seeds, fertiliser and labour, but excluding overheads. Use of a conversion factor of this nature requires detailed information gathered from a sample of single farms or standardised data over the whole of the area being considered. In the United Kingdom detailed cost information is available only for a small sample of farms amalgamated into larger administrative units to avoid disclosure problems. This method has many of the disadvantages of the standard man-day units but it uses more farm variables than labour alone. Adeemy[58] points out that this method is likely to over-emphasise those farms where there are high margins because they have low variable costs but high overheads.

3. Gross output data

In this case standardised values for gross output are used. Yield information is available on a county basis in the United Kingdom based on information collected from sample farms by the Ministry of Agriculture. Such gross yield data can be multiplied by the current price to achieve a gross output figure. Adeemy[59] chose a three-year period to smooth out yield and price changes.

TABLE 14 Major agricultural enterprises and their standard man-day equivalents, Acle parish, Norfolk, 1968

Crop acreage	%	*SMD equivalents*	%	*Livestock units*	%	*SMD equivalents*	%
Grass	54·2	913·9	5·7	Cattle, beef	75·1	1,177	26.6
Barley	20·9	1,412·0	8·9	Cattle, dairy*	13·8	2,500	56·5
Wheat	10·3	698·0	4·4	Pigs	6·8	321·5	7·3
Sugar beet	9·9	4,165·6	26·3	Poultry	4·3	426·75	9·6
Vegetables/glass	3·9	7,989·2	50·4	Sheep	0	0	
Small fruit/orchards	0·3	486·5	3·1				
Potatoes	0·2	112·0	0·7				
Stock feeding	0·2	61·75	0·4				
Others	0·1	3·5	0				
Total	100	15,842·4	100		100	4,425·2	100

*Dairy Cattle include only items 35, 37 and 39 on the June returns, that is: cattle and heifers in milk, cows in calf but not in milk and heifers in first calf, mainly for producing milk or rearing calves for the dairy herd.

(Based on data published by the Ministry of Agriculture, 1968)

There is little doubt that all three possibilities are beset with difficulties. Coppock[60] concludes that the balance of advantage lies with the use of information on labour as it can be applied to the smallest unit for which agricultural information is available and is also likely to change less rapidly than prices. He also concludes that inter-regional differences in efficiency will be compensated for, to some extent, in the United Kingdom by the fact that the western farms are basically livestock farms while those in the eastern half of the country are basically arable farms. Of the two money indices, gross margin data are the more valid as they represent the value added to the farm by each enterprise but, as they assume standard cost structures as well as average yields and prices, it is probably better to use gross output data.[61]

It is possible to avoid the problems of these standard indices. Thus, Jones[62] uses the acreage proportions of different crops, with the numbers of livestock per acre, to produce a classification. Similarly Scott[63] uses both crop combinations and livestock combinations to arrive at a composite agricultural regional breakdown of Tasmania. The boundaries of the consolidated arable and grassland groups are superimposed on those of the livestock regions. In the arable regions the basic cash crop features are then incorporated. The result is a classification of the land of Tasmania into nine classes.

Assuming that one of the three standardisation methods has been adopted, there are two basic approaches to the resulting data. The first is to map and regionalise on the basis of the leading enterprise and the second is to use Weaver's method of mapping associations, using in this case all enterprises on the farm as percentages of the total standard man-day units required.

Birch, after considering the problems with using parish data, surveyed a sample of 232 farms on the Isle of Man and produced a type of farming map by mapping the leading enterprise of the surveyed farms. Birch concludes:

> For a mixed farming landscape, such a map is limited in its value, since it ignores the important variations in subsidiary enterprises. But the dominant enterprise is a useful concept for it represents the farmer's strongest, if not his full, reaction to the physical and economic stimuli of his regional and farm environments.[64]

Napolitan and Brown[65] describe in some detail the division of England and Wales into six types of farming based on predominant farming enterprise. This work by the United Kingdom Ministry of Agriculture was based on a sample of individual holdings rather than on other, larger, units as there is no confidentiality problem involved for people working within the Ministry.

The major classification was into six types:

General arable cropping	A
Horticulture	H
Dairying	D
Beef cattle and sheep (livestock rearing and fattening)	L
Pigs and poultry	P
Mixed farming	M

If more than half the standard man-days were assigned to one of the first five types of production the farm was allocated to that class. All remaining farms which did not fall into one of the first five categories on this basis were classified as mixed. Maps of the distribution of the different types of farm, with the data amalgamated into ten-kilometre grid squares, are published by the Ministry of Agriculture.[66] Church[67] published a colour map showing the distribution of these different types of farms. Each ten-kilometre grid square is classified into one of thirteen types according to the estimated proportion of the total standard man-days in each square devoted to each type of production. Figure 21 shows a portion of this map redrawn in black and white.

Adeemy,[68] using published parish summaries for 1963 to 1965 for four counties in north Wales, produces a type of farming map where the predominant enterprises are mapped. He stresses that this is a classification of parishes rather than farms, but none the less he considers this a type of farming classification. Using gross output data instead of standard man-day units as a method of standardising the variables, he produces a threefold classification into cash crops, livestock and mixed farming. Those parishes which have less than 60 per cent of the total gross output derived from either cash cropping or livestock production were placed in the mixed farming category. The three main categories were then subdivided. For example, in the livestock category, parishes with more than 75 per cent of the total gross output derived from dairying were designated as predominantly dairying. If the total gross output from dairying lay between 75 per cent and 50 per cent the parish was categorised as mainly dairying. From this subdivision nine categories were derived in all. This classification was then compared with that produced for the same area by the Ministry of Agriculture using standard man-day data. The results were found to differ considerably. The reason for this difference was said to lie in the different approaches to the standardisation

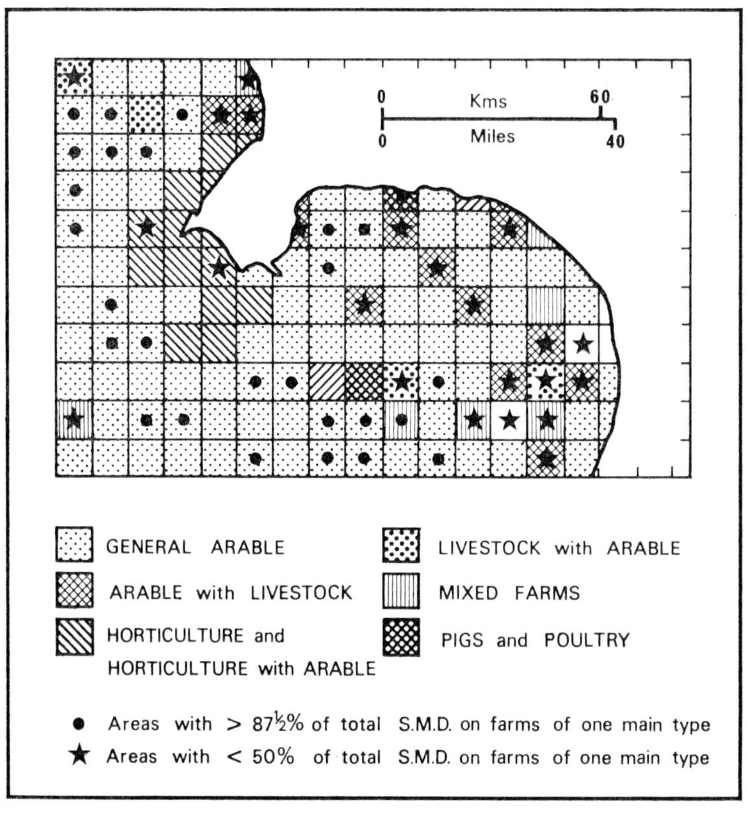

Fig 21. A portion of a farming type map of England and Wales based on agricultural census data for 1965. The area types are determined by the relative importance of different enterprises expressed as standard man-day requirements. (After Church, B. M., *et al.*, 'A type of farming map based on agricultural census data', *Outlook on agriculture*, 5 (1968), 191–6.)

of the data, Adeemy concluding that the method he adopted was more satisfactory.

Within each of the classes based on predominant enterprises, there is usually a considerable range of variation. Some of the problems involved are illustrated by Bennett-Jones[69] in a study of farming types in the East Midlands. The farms within each type of farming area[70] were classified according to economic criteria, Table 15.

TABLE 15 Percentage of farming types within farming areas, England and Wales, 1941

Types of farming Area group	*Economic classification type of farm groups**										*All*
	1	*2*	*3*	*4*	*5*	*6*	*7*	*8*	*9*	*10*	*groups*
A. Predominantly dairying	84	2	†	1	1	†	†	8	2	2	100
B. Dairying supplemented by other enterprises	63	3	4	—	—	1	6	23	—	—	100
C. Grazing and dairying	34	8	—	3	1	8	10	27	3	6	100
F. Mixed farming with substantial dairying side	45	16	1	2	1	5	7	12	2	9	100
H. General and mixed farming	14	17	—	9	3	22	15	18	—	2	100
J. Farming based largely on wheat and cattle	9	16	1	13	5	22	17	12	1	4	100
L. Mixed farming based on arable production	19	32	—	8	—	12	5	8	—	16	100
M. Mainly corn and sheep farming	7	8	1	16	6	32	12	11	3	4	100
N. Corn and sheep farming supplemented by cash crops	†	15	1	31	7	21	14	7	1	3	100
O. Mainly cash crop farming	2	7	†	44	20	15	5	3	1	3	100
X. Land of small agricultural value	63	—	—	—	—	—	—	37	—	—	100
Y. Marshes	15	19	—	13	3	20	12	12	5	1	100
All areas	33	10	1	12	4	13	9	12	2	4	100

* 1. Dairying
2. Cropping with dairying
3. Cropping with pigs or poultry
4. Predominantly arable
5. Predominantly arable with some livestock
6. Cropping with livestock of some importance
7. Cropping with livestock of considerable importance
8. Livestock
9. Poultry
10. Market gardens

†Less than 0·5 per cent

(Based on Ministry of Agriculture and land utilisation survey of Britain, *Type of farming map of England and Wales*, 1941)

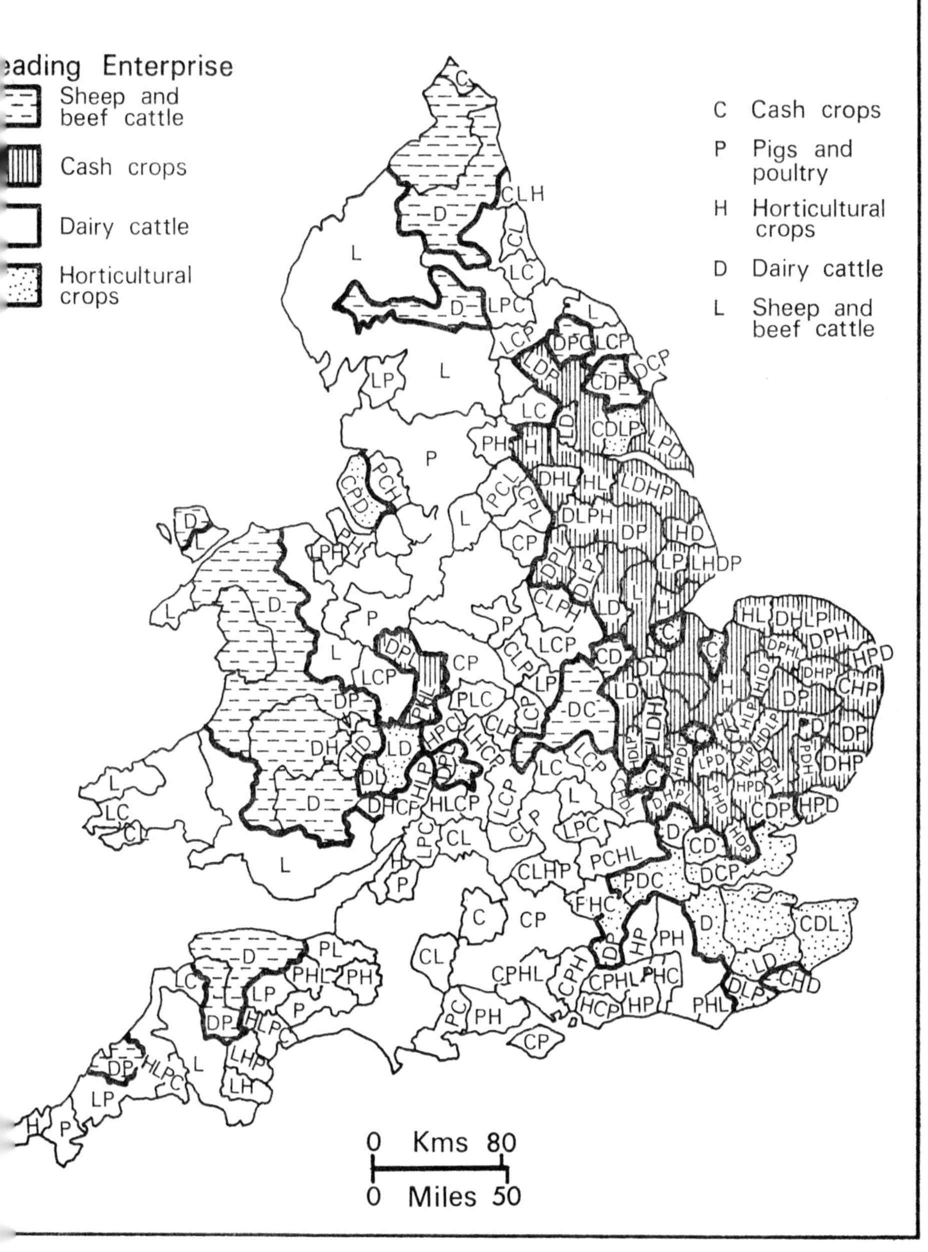

Fig 22. Enterprise combinations for the national agricultural advisory districts of England and Wales, 1958. The leading enterprises are shown by the shading and others in the combination systems by overprinted letters. (After Coppock, J. T., *An agricultural atlas of England and Wales*, 1964.)

In contrast to these classifications based on the predominant enterprise, Coppock[71] attempts to take up Weaver's point that it is not so much what is grown or produced that is important but the combinations in which it is produced that should be the distinguishing characteristics of type of farming areas. Using standard man-day material, each of the 350 National Agricultural Advisory Districts in England and Wales was allocated to a particular combination of enterprises in just the same way as has been illustrated for crop combinations (Figure 22). Parishes were not used as there are over 10,000 of these units in England and Wales and Coppock concludes that in some respects, despite their greater size, the NAA Districts are more satisfactory as they are of more uniform size than parishes. Seven enterprises were identified: dairy cattle, beef cattle, sheep, cash crops, fruit, vegetables, and pigs and poultry.

Symons[72] points out one of the problems brought about by the use of these large units. In the Lake District the large districts are so arranged as to include areas of upland rough grazing and lowland plain, resulting in the whole area being shown as dairying with livestock. Figure 23 shows this technique applied to the Republic of Ireland using information from the rural districts. The predominance of dairying within the farming of the Republic is clearly illustrated. The exception is a band running east-west through the centre of the country where beef and sheep are dominant. Even in this area dairying achieves secondary importance. In the area around Dublin the rearing of bloodstock, although not commonly thought of as a part of agriculture, reaches a level sufficiently high to appear in the combinations.

It is noticeable that in Acle parish (Table 16) the impression of the agriculture of the parish changes with the changes in the type of agricultural data used to analyse it.

The consideration, at some length, of efforts to regionalise variables related to agriculture illustrates both the diversity of the subject of agriculture and the importance which has been, and continues to be, placed on the idea of a region. The term region is familiar in everyday use. We all have some idea of the area defined by the 'mid-west' or by the 'east Midlands' and such terms provide convenient shorthand in discussion about features of the earth. On the other hand we must remain conscious of two difficulties. The first is that such regions must be carefully and unambiguously defined if they are to be useful in later investiga-

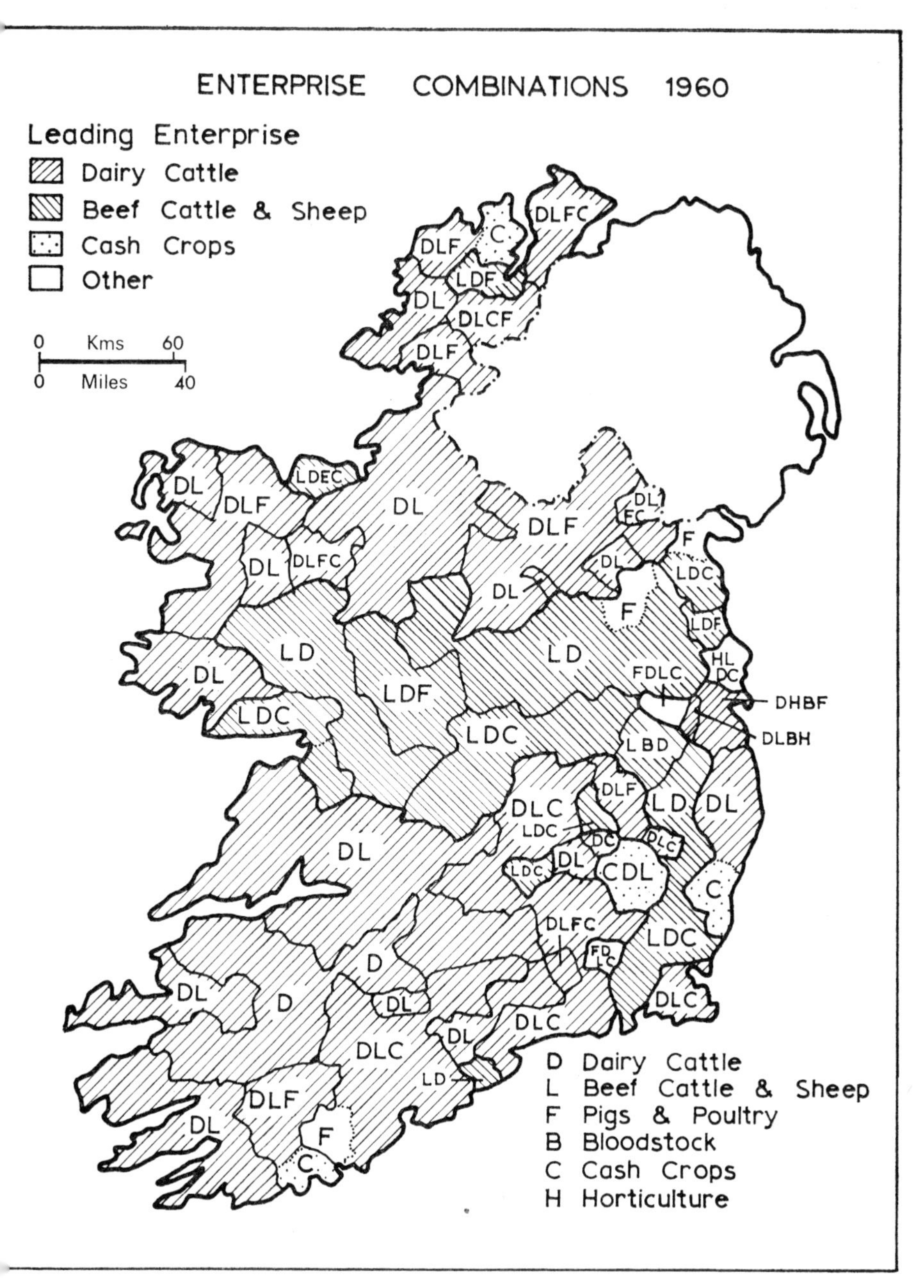

Fig 23. Enterprise combinations for the rural districts of the Republic of Ireland. Leading enterprises are indicated by shading and by overprinted letters.

TABLE 16 Enterprise combinations, Acle parish, Norfolk, 1968*

Enterprise	*SMD units*	%	*1-enterprise model*	D^2	*2-enterprise model*	D^2	*3-enterprise model*†	D^2	*4-enterprise model*	%
Vegetables	7,989·2	39·4	100	3,672·4	50	112·4	33·3	37·2	25	207·4
Cash crops	6,387·6	31·5	0	992·2	50	342·2	33·3	3·2	25	42·2
Dairy	2,500·0	12·3	0	151·3	0	151·3	33·3	441·0	25	161·3
Beef	1,197·0	5·8	0	33·6	0	33·6	0	33·6	25	368·6
Others	979·1	4·8	0	23·0	0	23·0	0	23·0	0	23·0
Pigs and poultry	748·2	3·7	0	13·7	0	13·7	0	13·7	0	13·7
Fruit	486·5	2·4	0	5·8	0	5·8	0	5·8	0	5·8
Sheep	0	0	0	0	0	0	0	0	0	
Total	20,287·6	100		4,892·0		682·0		557·5		822·0

*The enterprises selected here are those used by Coppock, J. T., *Agricultural atlas of England and Wales*, 1964.
†Vegetables, cash crops, dairy cattle.
(Based on data published by the Ministry of Agriculture, 1968)

tions. The second difficulty is that regionalisation must have some clear purpose if later work is not to be constrained within regional classes with little obvious meaning. Because the region is such an important feature in geography, especially perhaps in agricultural geography, further difficulties and alternative methods of approach will be examined in the next chapter.

CHAPTER FIVE

Problems and Alternatives in Regionalisation

In the previous chapter a comparison was drawn between classification and regionalisation in agriculture. There are two important problems to be overcome when using techniques of classification with agricultural distributions. The first, and most fundamental, is the identification of the individuals who are to be classified and the second is the identification of the discontinuities between sets of individuals which correspond to the boundaries between classes. Neither of these problems has received much attention in the study of agriculture.

THE PROBLEM OF THE INDIVIDUAL

By definition schemes of classification or regionalisation are concerned with the grouping of individuals. In some cases the identification and delimitation of these individuals will be relatively straightforward. If, for example, we are concerned with a 'type of farming' classification the farm can be taken as the basic unit and a classification devised accordingly. There may be some operational difficulties in the identification of cases where two or more farms are being operated as one unit, but the farm is usually a readily identifiable economic unit. Much more serious problems are presented by attempts to classify land, land use or agriculture. The identification of the basic individual becomes more difficult and is a question which has received scant treatment by geographers, despite the fact that it is of fundamental importance to the interpretation and value of the results of regionalisation and classification procedures.[1]

D. Harvey, too, has recognised this and states the problem very clearly:

> The regionalisation problem has one outstanding complexity which differentiates it from the grouping problem—if the observations comprise an areal sample or a sample of areas, and these observations are arbitrarily selected segments of space, it is well known that regionalisation resulting from the analysis is valid only for that particular segmentation of space which is used to define the units of observation.[2]

Thus the areal units adopted for classification must have some formal or functional justification for the classification to have any meaning.

Plants, animals or farms are discrete and separate entities and present few problems to the classifier. On the other hand the earth's surface is not divided into discrete and irreducible units. At the simplest level the earth's surface is a continuum which has to be divided into units which have some practical meaning. This problem is not unique to agriculture but affects all features of the earth which show some measure of continuous variation. It has received more attention in the study of soils and in ecology than it has in geography or agriculture.

Soil can be regarded as varying more or less continuously in three linear dimensions, within the limits set by the earth's surface and the top of the unweathered parent material. There are two distinct problems, the first of which is the demonstration of the existence of a soil individual and the second the demarcation of its physical extent. Soil profiles are used in the study of pedogenesis and therefore they tend also to be used for soil classification: Jones[3] concludes that, while they may be suitable for the former, they certainly are not suitable for the latter. The soil varies in three dimensions but to all intents and purposes the soil profile is a one-dimensional object recording how the soil at a point varies with depth. It is true that many soil profiles are from pits which have a three-dimensional extent. This is done in order to determine the character of the boundaries between the different soil horizons. None the less the information gathered from them is contracted to a single point for purposes of classification. Jones is led to the conclusion that, as classification must involve systematic organisation of individual units, it is doubtful whether any classification can be applicable to soils as they do not exist as circumscribed units. Alternatively the view is held that, while the soil individual is not found as a discrete entity, clearly separate from all others, it does grade *on the margins* into other soil individuals with unlike properties.[4] This implies that the major problem is not the identification of the individual but the definition of the boundaries between individuals: the existence of a series of recognisable types is assumed. The concept of the soil individual must be explained before classification can proceed. In pedology the smallest recognised unit is called, appropriately, a pedon. This is an area between one and ten metres square, its size depending on the variability of the soil being considered, and is conventionally regarded as roughly hexagonal in shape. If soil properties simply varied in all directions from a particular spot the ideal unit within which there is some degree of homogeneity of variables would be a circle, but the nearest geometric shape to a circle which allows no overlap and leaves no areas outside the unit is a hexagon.

When classifying land a similar component of landscape is required. This can be defined as an area within which there is, for all practical purposes, no variation in climate, physiography, geology, soil and edaphic factors.[5] Most later workers on land classification have allowed some degree of internal variation within the basic land unit, often called a *facet*.[6] The important point is that this internal variation should show no sharp breaks within the facet. A similar unit called a *facies* has been defined by Russian landscape analysis.[7] In all cases there is an assumption that the homogeneous element of the landscape which could effectively form the basic unit of classification is as arbitrary as the class limits that define it. The choice of boundary lines between facies or facets becomes a matter of judgement and experience.

Anderson[8] recognises two views of vegetation studies: the first regards plant communities as complex organisms where plants growing together sufficiently modify the environment to form a recognisable and repeatable vegetation grouping; the second regards no two communities of plants as strictly the same since there is a continuous variation in their detailed composition. The former view involves classification and the latter ordination, that is the placing of a particular community or plant association on a continuum of possible communities. Lambert and Dale[9] also recognise this distinction but it is clear that it is the *site* that is the individual which is being classified or placed on a continuum. The site is conventionally a sample quadrat, the size and the distribution of which is determined by the investigator taking into account the complexity of the plant associations with which he is concerned. The site, then, is not a clearly defined individual which is to be grouped in ecological studies any more than the pedon is in pedology. The quadrat is an operationally defined unit and it is on the size and the spacing of these units that the resulting classification or ordination will depend.

Further parallels may be seen in the case of climatic regions in which the individuals being classified are meteorological stations at which data are collected. The distribution of these may be far from representative of the world's climates, though the World Meteorological Organisation has recommended that there should be one major station at least every 600 nautical miles. Clearly weather and climate vary constantly both horizontally and, within certain limits, vertically. The individuals being classified to form the climatic regions are theoretically infinitely small points on the ground. In practice operational units are chosen depending on the scale of the analysis and the frequency of the available or the assumed data. In the face of these difficulties and often without a complete understanding of them geographers have resorted to operationally defined units. These are usually administrative areas such as parishes or counties, or, more rarely, grid squares of various sizes. It is clear that the farm has some practical meaning as a unit. This is not true of parishes or counties or even grid squares.

The difficulties concerning the availability of information discussed in Chapter 3 provide part of the explanation of the use of statistical units other than farms for classification purposes. It is clear that no large-scale regionalisation of agriculture could take place using the farm as the basic unit. It is necessary, however, to consider carefully what the results of a regional scheme of agriculture imply when it is based on operational units rather than on other functional units.

One further example will serve to reinforce this important point. Zobler,[10] in testing the significance of regional boundaries, defined nine regions in Salem County, New Jersey, using geological and physiographic variables. Information for soils, land use, type of farming and population structure were then collected for each of the nine regions. Using the Chi square test the significance of these regional boundaries for the latter four variables was considered. The test is derived from the equation for the Chi square statistic:

$$\sum_{i=1}^{n} \frac{(o_i - e_i)^2}{e_i}$$

where o_i equals the observed frequency of the variable in the ith region and e_i is the expected frequency in the ith region. The expected value is calculated assuming that the regional boundaries hold no significance and all regions are the same. Some of the observed and expected frequencies used in this analysis were measured in acres. Thus Zobler regarded an acre as the individual unit of land to be classified. Mackay[11] has pointed out that had Zobler used larger units, such as square miles, the differences between the expected and the observed frequencies would have been considerably smaller and the size and the significance of the Chi square statistic would have been reduced. The significance of Zobler's results was severely affected by the size of the unit chosen. Mackay continues:

> If Zobler could demonstrate that in his study there was a theoretical basis for the selection of unit areas of one acre, then the above criticism would obviously not apply, because the frequencies would be absolute not relative.

While presenting arguments for the use of acres rather than other units of measurement, Zobler[12] was unable to show that acres have anything other than an operational value as individuals. Classification or regionalisation without a clear understanding of the nature and the validity of the individual can be misleading and this problem warrants further attention in geographical writing.

REGIONS OR CONTINUOUS DISTRIBUTIONS?

The problem of breaking down a distribution into regional classes becomes one of identifying regional boundaries. It is not difficult to divide an area into regions, with any amount of arbitrary choice of boundaries, if the resulting regions are justifiably convenient units in which to study the complex interrelationships between other geographical phenomena. The problem is not to be found here: it is rather in the regional division of land on the basis of one or more phenomena as a description of that distribution or distributions and as a step towards the attempted explanation of that distribution.

It has long been realised that regional boundaries drawn on maps indicate only zones of change. But such change is apparent over all points of the earth to a greater or lesser extent. The boundaries, therefore, tend to be drawn where one characteristic becomes dominant over another. It is commonly recognised that economic regions have cores of maximum specialisation which decreases outwards. The fact that the margin is influenced by more than one economy makes the boundary areas more complex than the cores. Baker's[13] interest lay mainly in the core areas of agricultural regions but some of his divisions are marginal, multi-product regions separating single-product zones.[14] Other descriptions of agricultural or land-use regions stress their characteristic of gradual change. H. H. McCarty[15] recognises three characteristics of agricultural regions: core areas, mixed boundary zones, and impermanent boundaries. Such points are reinforced by Kimble[16] who stresses how, for example, the wheat belt of the American prairies merges almost imperceptibly into the hay and dairy belt to the east and into the corn belt to the south. From the air, at least, it is the links and continuities in the landscape that are important rather than the breaks.

This recognition of a gradual shading from one land-use core to another has led to some special emphasis being placed on the characteristics of land use and agriculture of marginal areas. Probably the first person to consider the margin worthy of special study was Johann Heinrich von Thünen (see Chapter 2). The fluctuations of the economic margin between rent and no-rent, especially with such factors as farmers' income, is a case in point. From this concept of a fluctuating margin of production has stemmed a considerable literature.[17] These economic studies need not be restricted to the actual limit of all production but each of von Thünen's zones showed decreasing returns from a particular crop or enterprise with increasing distance from the market, until, eventually, the crop or enterprise is replaced by another. A detailed analysis of such a marginal area is provided by Gibson[18] who defined a cheese region of the dairy belt of the United States and a meat-animal region of the corn belt identifying these as counties with at least 45 per cent of the total farm income coming from one of these sources. Lafayette County, lying geographically between these two areas, and having both cheese and meat-animal income, but neither to the specified 45 per cent level, is taken as representative of the margin where the two agriculture systems merge.

So far we have considered only those studies where regional boundaries have been defined. There has been no attempt to test the validity of these regions statistically. There is no evidence that the use of such terms as the corn belt is accurate or inaccurate as related to the actual distribution of agriculture. But once regional boundaries have been defined it is possible to test the degree of difference between the regions in terms of a number of variables.[19] The simplest way to do this is to use the Chi square test. But, whereas this technique can assess the significance of a set of regional boundaries, it gives us no indication of the alternative groupings of the basic units, nor, therefore, does it assess the relative significance of the regional boundaries against all other such divisions of the total area, including the possibility of considering the whole area as just one region.

To illustrate this an area is divided into sixteen grid squares in each of which is a figure representing the number of farms in the grid square (Figure 24). In the first example these grid squares are divided into four regions on the basis of the number of farms per unit area. In region A the observed frequency of farms is eleven and the expected frequency is $2 \times 304/16 = 38$, which would be the number of farms in the region if they were evenly spread over all the regions. Repeating this operation for the other three regions yields a Chi square value significant at the 99·9 per cent confidence level. These regional boundaries proposed are highly significant in terms of the distribution of farms within the area. We next examine an alternative regional combination of the basic grid squares so that region A now occupies the north-western four cells, region B the south-western four cells, region

C the north-eastern four cells and region D the south-eastern four cells. The observed frequency in region A is now 28 and the expected frequency is 76. In this case the Chi square value is also significant at the 99·9 per cent level. This shows that a completely different and highly significant regional division is possible.

There are many other ways of creating regional boundaries, some of which have been dealt with in the last chapter while some will be introduced in the next. Whatever the technique applied, the data for a number of individuals is always combined into a number of regional units. Different combinations of the

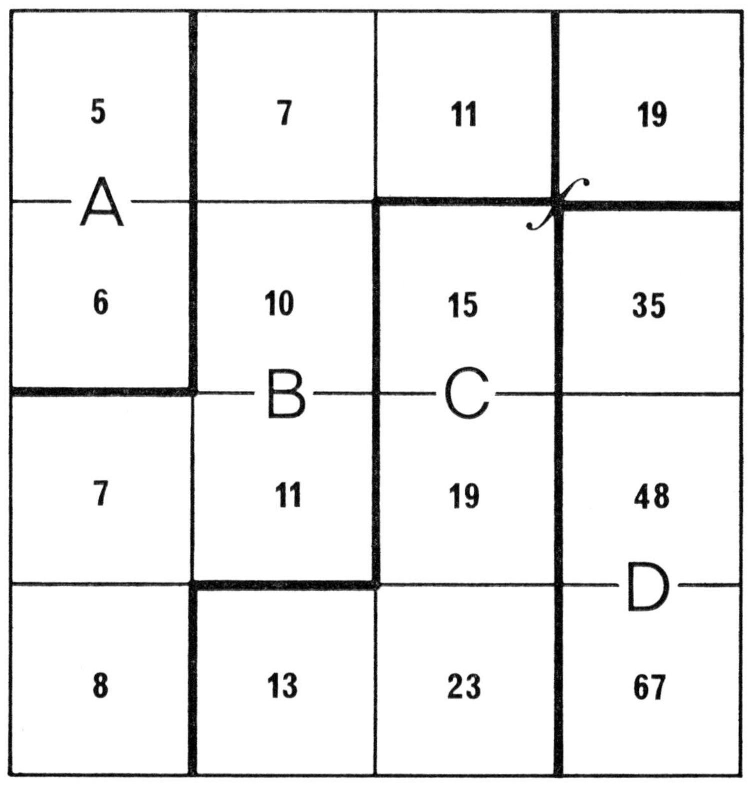

Fig 24. Possible regional divisions of 16 grid squares on the basis of density of farms per unit area represented by the number in each cell.

basic unit can be tested to show the best possible combination and therefore the most discrete regions, but, once the regionalisation process has been started, the result will always be a set of regions. There are no tests to show whether the regions themselves are the best description of the data used. They are assumed to exist, are searched for and found, even if there are no recognisable breaks on the ground.

The problem of defining regional boundaries is really a two-dimensional representation of a similar problem in the study of settlement. After Walter Christaller[20] expounded central place theory for the first time, and indeed for some time before his work, many geographers and economists were concerned with the possibility of the existence of a hierarchy of settlement. After Christaller postulated this hierarchy theoretically and geometrically and later workers illustrated its characteristics in empirical studies, later attempts were made to verify it.[21] Could settlements be divided into hierarchical classes, each break or boundary clearly and irrevocably established? Vining[22] held the view that settlements could, in this respect, be likened to puddles, ponds and lakes: we do not know when a puddle is large enough to be regarded as a pond, nor when a pond is large enough to be called a lake. The three divisions are merely satisfactory working units and have very little validity. This description can often be applied to regional boundaries including agricultural regions.

ISOPLETH MAPS

When geographers wish to represent the pattern of a continuous distribution, such as height above sea level or barometric pressure, iso-lines connecting places of equal height or equal pressure are drawn. The result is a two-dimensional representation of a continuous surface. Except in a very general sense we do not talk of regions of pressure or of regions of height. It is only when many other variables are added that the configuration becomes sufficiently complex to lead to the creation of climatic regions designed to show a similarity of interaction between these variables over a particular area. The position is in many ways similar in agriculture. The distribution of one, or perhaps the relationship between two, agricultural variables can be represented with an isopleth map. This can be used to emphasise the essential continuity of the distribution concerned. There are no clearly defined boundaries in an isopleth map, just a gradual shading off in the importance of the variable.

There is one immediate difference between isobars, contours and isohyets on the one hand and the isopleths of agricultural variables on the other. The former are plotted from point data, the information is gathered at theoretically infinitely small points. Agricultural variables and ratios are collected over finite areas and would normally be meaningless at units smaller than the farm. Further, statistics are usually available only for the large administrative units. In such circumstances it is normal to draw choropleth maps, shading each administrative unit according to the importance of the variable concerned within it. Alternatively the administrative units can be 'squared off' and in Figure 25 some of the parishes of Norfolk have been allocated to one-kilometre grid squares. The area of each parish is calculated to the nearest square kilometre and each is then allocated the requisite number of squares. In order to fit these together some of the parishes may be displaced somewhat so errors of location are added to small errors of area. These squares, or larger ones, can then be used to represent the distribution of variables (Figure 26) with a blurring of the apparent abruptness at the edges of administrative units. It is also possible to place a point in the centre of an administrative area and to take this point as representative of the area, applying the value obtained from the whole area to that point.[23] Isorithms are drawn from data which are available at points while isopleths are drawn from data representing ratios collected from areas and contracted to representative points in these areas.

The selection of the control points for each unit of area has some pitfalls. One is the selection of the centre of an irregular shape. Although some work has been completed on the mathematical description of shape[24] and concerning the centre of point patterns,[25] little or nothing has been written about the definition of the centre of an irregular area.

The definition of the centre of gravity of an irregular area is represented in Figure 27. The first step is to 'square off' the boundaries of the area into a series of straight line segments. The resulting shape is then broken down into a number of triangles. There is no unique solution to this operation but different methods of producing the triangles have no effect on the result. The centre of gravity of each triangle is then found, t_1, t_2, t_3, . . . t_i. These are located at the intersection of the three bisectors of the internal

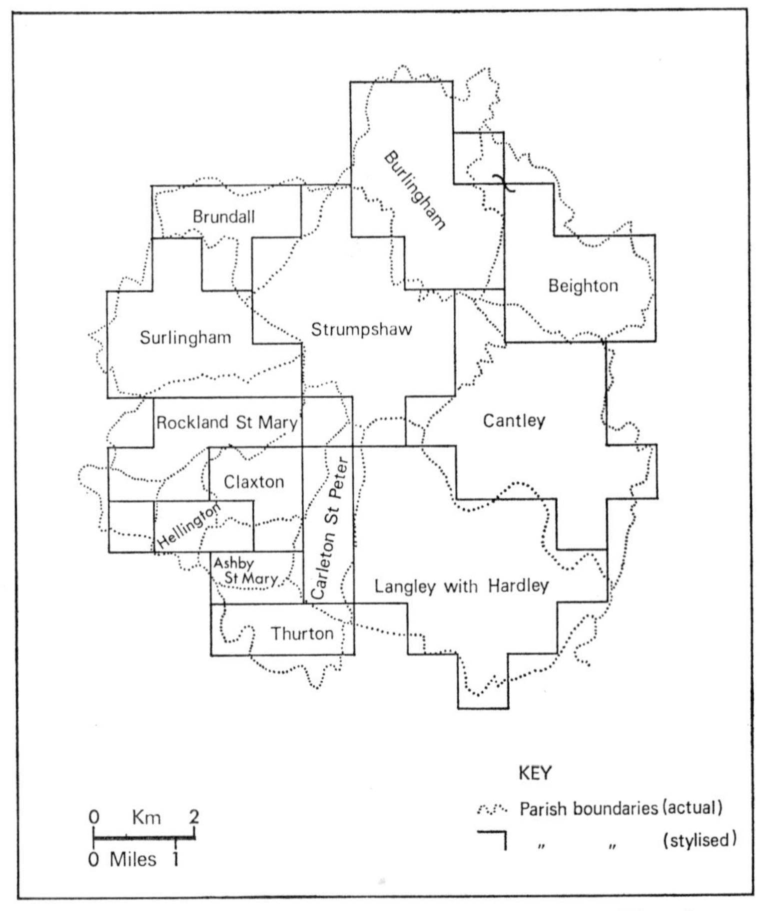

Fig 25. A group of 13 Norfolk parishes assigned to km grid squares. This assignment has taken place while retaining, as far as possible, the correct areas and relative locations of parishes.

angles of the triangle. A rectangular co-ordinate system is then established around the shape; the orientation of the x and y axes is irrelevant. A line is drawn through approximately the centre of the shape, parallel to the x axis. The centre of gravity, D, along this line is given by:

$$Dx(m_1+m_2+m_3+\ldots) = m_1x_1+m_2x_2+m_3x_3+\ldots$$

where m_i is the area of the ith triangle, x_i is the distance of the centre of gravity of the ith triangle from the y axis and Dx is the distance from D to the y axis. A second line is now drawn, at right angles to the first, passing through the point D. The process is repeated for this new line, and the new point D now yields the centre of gravity of the area.

There are, however, many occasions when the centre of gravity is not an appropriate point of representation of the area. The most extreme case is a shape with its centre of gravity outside its boundaries, for example, a horseshoe. There are other circumstances where the control point to represent the unit area can be arrived at with an element of subjective choice.

Using ten-kilometre grid squares for mapping, and selecting the centres of each of these grid squares to form the control points for the isopleths, means that the 'mesh size' of this data-collecting system is 100 square kilometres. No variation on a smaller scale will be represented on the resulting isopleth map. The ten-kilometre squares are assumed to be internally uniform. This is exactly the same as using the grid squares as individuals

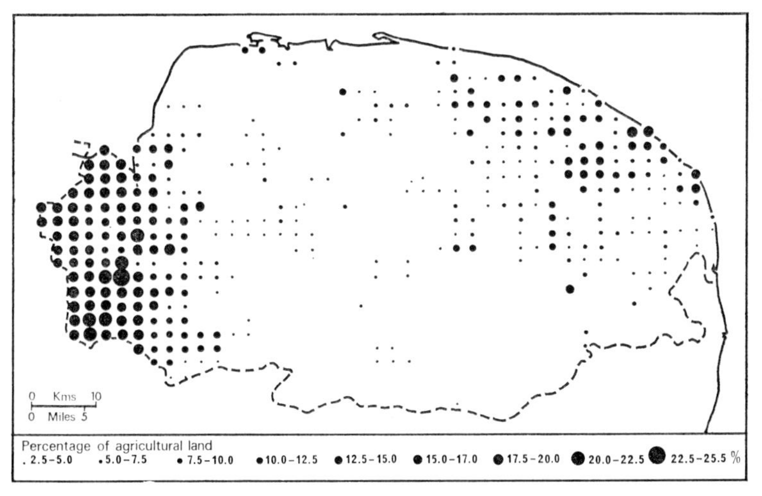

Fig 26. Parishes have been allocated to 2·5 km square grid cells. Where more than one parish had land in one cell that parish having the greatest proportion of its land area in the cell was allocated to it. The proportion of the agricultural land used for growing potatoes is indicated by symbols. As this map was originally produced by a computer line printer, the scale is only approximate.

to combine into larger units in a regionalisation scheme. Where one half of a grid square is known to contain only mountain land with no farms and where, for example, the proportion of the total arable land devoted to wheat is being plotted, it seems wrong to apply the same weighting to the part of the square that has no farmland as to the part of the square that is all arable land. The control point of the square would seem to be better placed in the centre of the arable area of the square rather than at the square's centre of gravity. Such subjectivity is

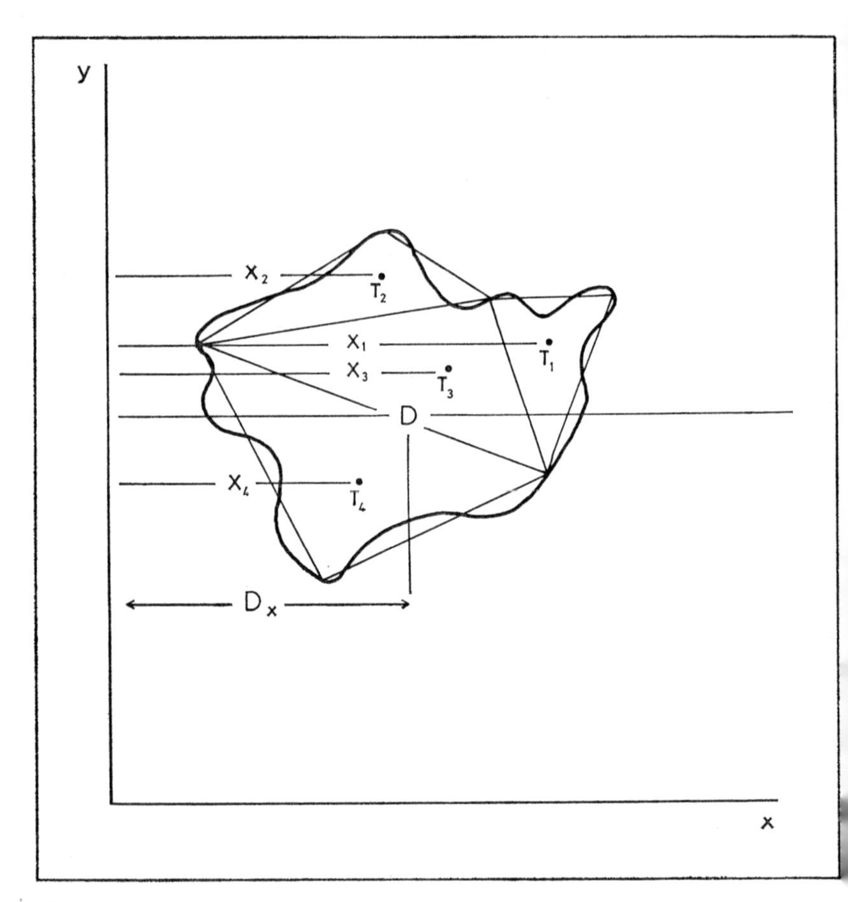

Fig 27. Method of finding the centre of an irregular shape.

possible only if it can be applied equally to all the grid squares. It is also necessary for the arable land within the grid square to have a recognisable centre. For example, in the peninsular administrative districts of south-west Ireland arable land tends to be distributed around the edges of the peninsulas, the centre of which is made up of rough mountain land.[26] The central point of the districts, or the central point of the arable land only, fall in uninhabited mountain land.

In the circumstances, and particularly taking into account the complexities of accurately defining the central points of areas of irregular shape and the fact that internal differences in the data areas are lost through the use of that particular mesh size, it is often best to define the centre of each area by 'eyeball' methods. When the control points are close together the placing of one point becomes relatively insignificant; where there is only a thin scatter of points the misplacing of one could be critical to the results.

PROBLEMS OF DRAWING ISOPLETHS

The problems of observational bias and sampling error in the use of isopleths has been developed at some length by Blumenstock[27] and need not concern us here. Although Blumenstock talks of 'perfect isotherm fits', in the concluding paragraph he remarks that it is not always straightforward to contour irregularly spaced point data. He pays little attention to this type of problem.

There are two sources of difficulty, interpolation and the selection of class intervals. To be able to interpolate the position of a contour line between two known points we need some hypothesis about the nature of the gradient between these two points. Is it linear, concave or convex? Where there is no evidence to the contrary, as is the case with most isopleth maps, it is normal to adopt the simplest hypothesis that there is a simple linear relationship. Interpolation can then proceed as follows (Figure 28A): at four known data points A, B, C, and D we have values of seventeen, fourteen, thirteen and thirteen respectively. A straight line between A and B is divided into three equal parts. The isopleth of value fifteen will lie two-thirds of the way from A to B. Similarly it will lie half-way between A and D and between A and C. This isopleth can now be drawn.[28] Over a more extensive area this is known as the polyhedron surface fitting technique. All points are linked together into a triangular lattice so that no point lies outside or inside a triangle.[29] This is the simplest method of isopleth interpolation. Not only does it assume a simple linear slope between the control points, but also there is no unique method of linking irregular points into a triangular lattice and there are, therefore, several contrasting solutions to the final isopleth plot. J. R. Mackay[30] points out the problem of the existence of cols to be resolved in interpolating isopleth patterns. If the control points are arranged on a rectangular grid serious ambiguities can arise (Figure 28B). In this figure two contrasting patterns can be chosen for the twenty value isopleth. One solution provides a col between the north-east and the south-west corners while the other reverses this by providing a ridge connecting the north-east and the south-west. Mackay suggests that this can be resolved by calculating the average of the two interpolated values at the centre of the square (Figure 28C) or by the use of the

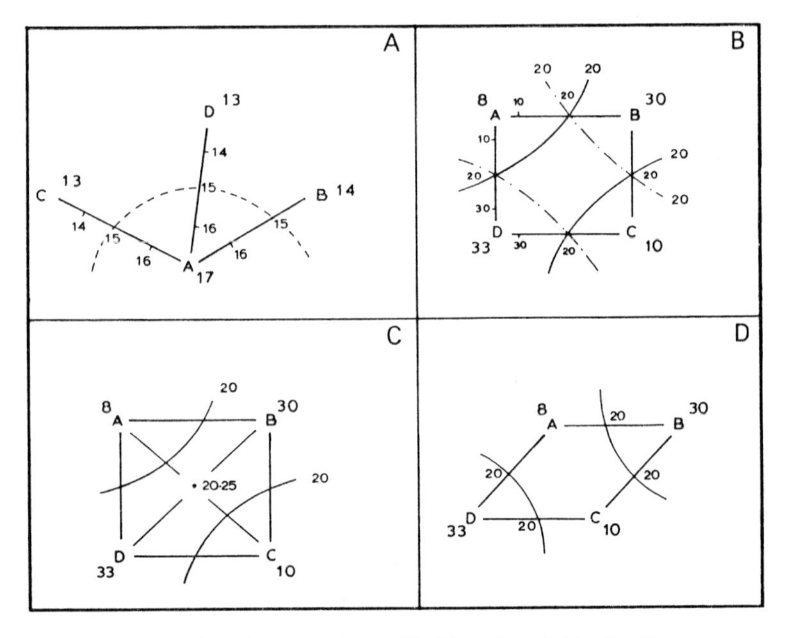

Fig 28. Methods of resolving ambiguities when plotting isopleths.

triangular lattice technique (Figure 28D). Extending this lattice technique will permit interpolation between each point and its six nearest neighbours.

However, ambiguities still exist in this pattern if interpolation is allowed between each point and others at irregular distances from it. Robinson and Sale[31] also show how, in certain circumstances, an artificial gradient can be produced by the use of regular rectangular units of area. We shall return to consider more sophisticated methods of isopleth interpolation after a brief consideration of the problems of the selection of class intervals.

The choice of class interval can have a critical effect on the nature of the resulting surface.[32] A variety of class intervals can be chosen,[33] but isopleth maps have one important difference from choropleth maps for which a similar variety of class intervals can be chosen. The closer together the 'contours', the steeper the gradient. This well-known property is true only if regular class intervals are chosen. It is perhaps odd, therefore, that Mackay[34] has found that the use of equal steps is the exception in isopleth mapping. Unequal steps can be used to emphasise the changes at either the upper or the lower end of the scale of variation. This is often of particular importance in highly skewed distributions where there are a large number of values between ten and a hundred and a few between a hundred and a thousand. A logarithmic class interval might be chosen but the significance of the isopleths will not be easy to interpret. Normally it is simpler and more consistent to use regular class intervals.

FURTHER SURFACE FITTING PROCEDURES

There are two more complicated techniques for interpolating values for the spaces between control points. The first is a series of procedures which can be grouped under the heading of polynomial surface fitting. The second involves numerical approximation over a uniform grid.[35]

In two dimensions a smooth curve can be fitted through a series of points by an equation of the form:

1. $z = a + bx + cx^2 \ldots nx^i$.

The more complex the curve the greater the number of terms required in the equation. For a simple straight line the equation is:

2. $z = a + bx$.

The same principle can be applied in three dimensions by adding a further independent variable. The equation for a simple plane surface becomes:

3. $z = a + bx + cy.$

More complex surfaces require more terms in the equation which takes the general form:

4. $z = a + bx + cy + dx^2 + exy + fy^2 + gx^3 + hx^2y + ixy^2 + jy^3 \ldots$

Theoretically a perfect fit can be obtained to any collection of control points. Unfortunately this final fit may have no validity in the areas between the control points even though the surface passes through all the points. Because of the nature of a polynomial surface, patterns established in one area will be repeated elsewhere, even if there are no control points to justify this. Lower degree surfaces, that is surfaces which are defined by equations with few terms in them, may be used to isolate large-scale trends in the data while not passing exactly through each of the control points. This use of what is called trend surface analysis will be considered later as it is not concerned directly with fitting isopleths to data.

Because of the creation of spurious complexity in the surface fitted by a polynomial a modification has been proposed by Cole.[36] First a quadratic surface is fitted to the data points. This is defined from the equation

5. $z = a + bx + cy + dx^2 + exy + fy^2.$

Although this surface may not fit very closely to the data points, it will not provide complexities to the surface between the control points. Bumps above and below this quadratic surface are now fitted so that it is 'bent' towards the known values. This is achieved by computing the values at the intersections of a grid covering the area of the quadratic surface. The grid intersection A, nearest a particular control point, is identified. The computed values for the quadratic surface at the nine grid intersections around this point A are used, along with the known value at the control point, to fit a further, local quadratic surface. In this second case the known data value is given a weighting of four against one for the other nine points used to compute the second surface. This allows the known value to have four times the influence on the resulting surface as any of the computed values. This local surface is used to create a new set of computed values at the nine grid intersections immediately surrounding the control point. This process is repeated for each control point. If the resulting surface for this complete operation is still removed from the actual data values the whole process can be repeated as many times as necessary. The result is a surface which fits the known data and yet does not have undue complexities in the area between the control points. This technique may be of special use where we have agricultural information for small selected areas which are of an uneven distribution. A series of isarithms drawn through these points should not show undue complications in areas where the points are sparse. Clearly a contouring process of this complexity can be achieved only through the use of a suitable computer program.

Numerical approximation over a grid is a further attempt to avoid the difficulties involved in fitting a polynomial surface through a series of control points. First a regular grid is placed over the area. The grid spacing should not be more than one-third of the average distance between the control points. Values are calculated for each of the grid intersections in two stages:

1. Values are assigned to the corners of each cell which has one or more data values within it. The centroid of the data points within the square is determined and a surface is calculated to pass through this centroid and taking account only of the immediately surrounding data points. Each octant around the centroid is searched for the nearest control point and these eight points are used in the calculation of the local surface. The influence of each of the eight points on the surface is calculated according to its distance from the centroid, this influence usually falling off with the square of the distance. This process is repeated for all grid cells which have control points within them.

2. These established values are used to define the values of the grid intersections of cells in which there are no control points. Each established mesh point after stage one is used as the centre for calculating the unknowns. In this way the process continues outwards from the known areas into the less well-known and the surface thereby remains fairly simple between the control points.

Two things are clear from the foregoing discussion of isopleth drawing. First, contouring of all types is a subjective business and is potentially open to considerable operator bias; secondly, to carry out extensive contouring by any other method than by eye and to produce a map that is replicable, the large volume of mathematical interpolation requires the assistance of a computer. There are several contouring computer programs available, perhaps the best known of which is SYMAP, produced by the Harvard Laboratory for Computer Graphics.[37] In addition, the manufacturers of the various machine plotting devices have programs for this sort of operation and also programs for the drawing of three-dimensional representations of surfaces of all kinds. Machine-drawn isopleths are not necessarily more correct than those drawn by hand. The machine can operate only with the algorithm that it is programmed to follow. The advantage of machine mapping, apart from its relative speed, is that the maps drawn are exactly repeatable and in a series maps will all have the same built-in assumptions and bias.

When we are concerned with the distribution of a single variable, for example corn in the American Mid-West, it is probably better despite its difficulties to describe this distribution by a contoured surface than by demarcating a region within which corn growing is in some way characteristic.

THE USE OF TREND SURFACE ANALYSIS

The isopleth maps so far considered may be complex in pattern since the spatial variations in the observations are complex. As a result the surface may be difficult to interpret and the source of further hypotheses concerning the distribution may be obscured by too much detail. It is often desirable to simplify the surface in some way. One such way is to apply a moving average over the area.[38] A grid covers the area and a circle of predetermined radius is centred on each of the intersections of the grid in turn. The values of the control points that fall within the circle are averaged and this average is applied to the grid intersection currently forming the centre of the circle. This is repeated for each of the mesh points. The process is very similar to the use of a linear moving average except that in this case it is applied two-dimensionally. The use of a moving average is laborious, especially with a fine grid, and produces widely varying results with different sizes of circle and different sizes of the basic grid.

A more useful, more widely used and rigorous technique is provided by the application of trend surface analysis, though this has been little used in the study of agriculture.[39] Trend surface analysis has been widely used in geology[40] and geography[41] and in other subjects concerned with the spatial distribution of phenomena.[42] In all cases the object has been to separate large-scale gradients in a particular variable from smaller, local, gradients. For example, the distribution of types of cattle in the Republic of Ireland is normally described by distinguishing a dairy region centred on County Limerick in the south-west and a beef-fattening area centred on County Meath in the north-east. The beef calves of the dairy herd are reared in south-west Ireland and move north-eastwards to be fattened in County Meath and sold, through the port of Dublin, for export to the United Kingdom.[43] From the nature of this movement we would expect the change in emphasis from dairy cattle to beef cattle to be gradual and not to conform to a single regional boundary. The isopleth map* showing the distribution of dairy cattle as a percentage of total cattle numbers (Figure 29A) shows a confused picture, though the proportion of dairy cattle is highest in the south-west and lowest in the north-eastern counties. If we are to test the hypothesis that the cattle distribution reflects this internal movement of cattle we need to simplify this picture. If, at each of these control points, we place a thin rod, the height of which is in proportion to the percentage of dairy cattle in that district, we have a representation of the statistical surface which we wish to simplify. The simplest trend surface which we can fit to these data is a plane with a constant dip and strike. This surface will be defined by an equation:

$$z = a + bx + cy.$$

This plane will be a least-squares fit in the same way as a two-dimensional regression line is a least-squares fit through a scatter of points. The linear surface, therefore, slices through these rods that we have placed over the control points, in such a way as to leave some above the fitted surface and some that do not reach it. The least-squares criterion implies that the sum of the squares

* The isopleths have been drawn from control points representing the central points of the 158 rural districts in the Republic.

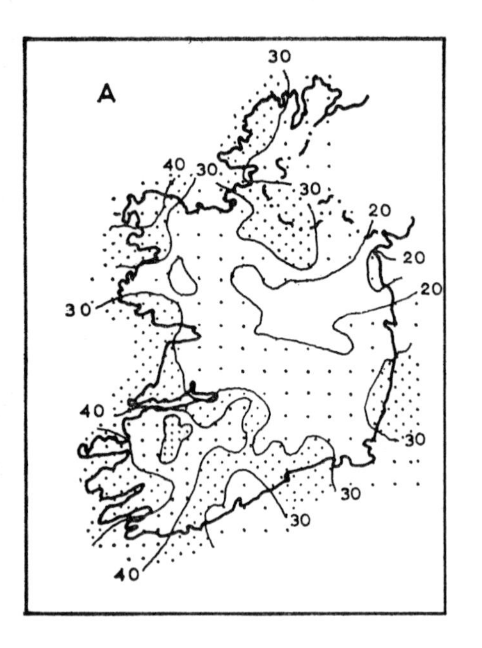

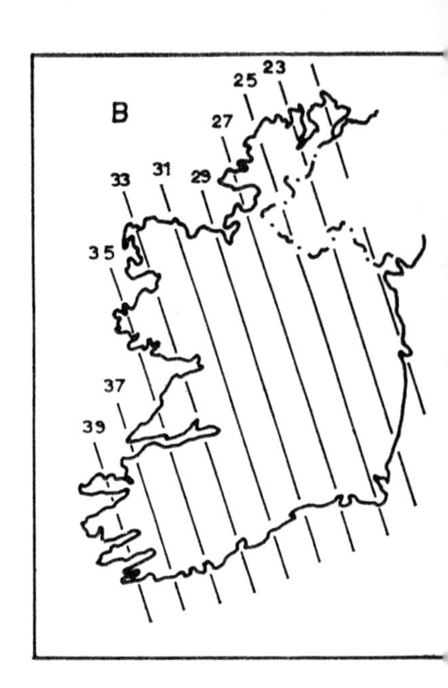

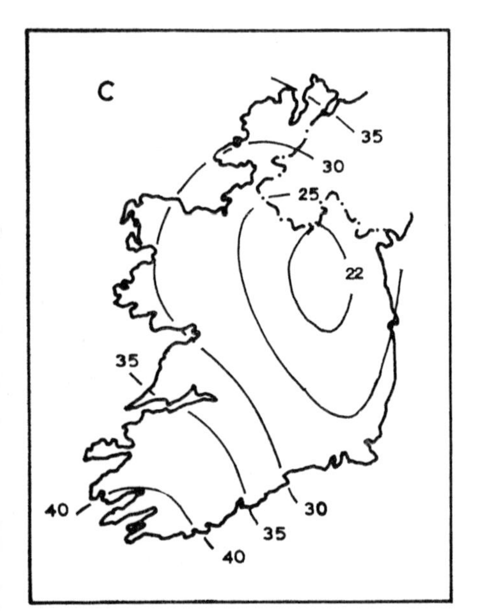

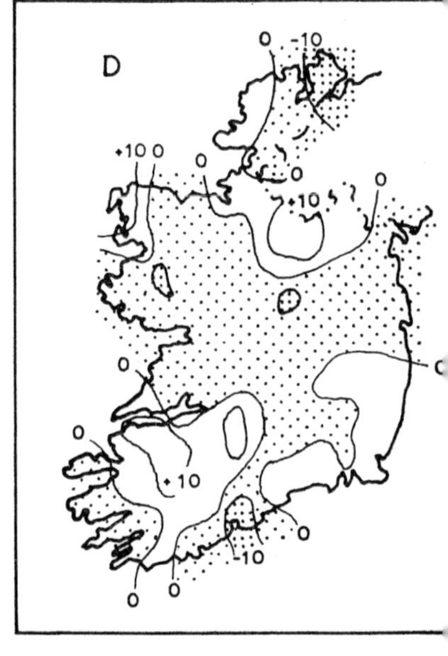

Fig 29. (*a*) Percentage of total cattle which are dairy cattle in the rural districts of the Republic of Ireland, 1961. (*b*) Linear trend surface of the percentage of dairy cattle in the Republic of Ireland. This surface explains 32·64 per cent of the overall variation in the distribution of dairy cattle. (*c*) The cubic trend surface for the distribution of dairy cattle in the Republic of Ireland. This surface explains 45·26 per cent of the variation in the distribution of dairy cattle. (*d*) Residuals from the cubic trend surface for the distribution of dairy cattle in the Republic of Ireland. (Source: Irish Ministry of Agriculture statistics.)

of the distances from the ends of the rods to the surface is kept at a minimum; the plane is, therefore, a best-fit surface to the data values. In the case of the distribution of cattle in Ireland this linear surface shows a clear trend in the proportion of dairy cattle from north-east to south-west (Figure 29B). The 'explanation' of this surface can be calculated in the same way as is possible with a regression line.[44] This linear surface accounts for 33 per cent of the total variation shown by the control points. A better fit to these values can be achieved by using a cubic surface, while this still retains the function of dividing the variation into local and large-scale components. The cubic surface shows more clearly the concentration of beef cattle in the area north and east of Dublin (Figure 29C).

A map of the residuals has considerable value also and shows the deviations between the actual values and the computed values for each of the control points (Figure 29D). It can be used to show the areas, and the extent, of local-scale variations from the overall trend. Clearly the simple cubic surface fails to take into account the heavy concentration of dairy cattle in the area of County Limerick and this remains therefore one of high positive residual values. Also, west County Mayo and Counties Cavan, Monaghan and Leitrim in the north-central part of the country have more dairying than expected. These are counties where the small farm predominates, too small to support more than a limited number of cattle. As dairying is a more intensive enterprise than other forms of agriculture based on cattle, a farm with only a few cattle is most likely to concentrate on dairying.[45]

If we use smaller data units there will be more local variation* in the data and the level of explanation achieved by a low order, or relatively simple, surface will not be particularly high. Using approximately 500 parishes in the county of Norfolk as the basic data units, a cubic surface of the distribution of cereals (Figure 30) shows clearly the falling emphasis on cereal production both to the east and to the west of the county as vegetable and fruit growing achieve more importance in the fenland of west Norfolk and permanent pasture begins to dominate in the river valleys of east Norfolk. This surface accounts for 34 per cent of the variation in cereal production in the county.

From these examples it can be seen that trend surface analysis

* The local variation can be considered as 'noise' which is obscuring the 'signal' we are trying to extract from the data.

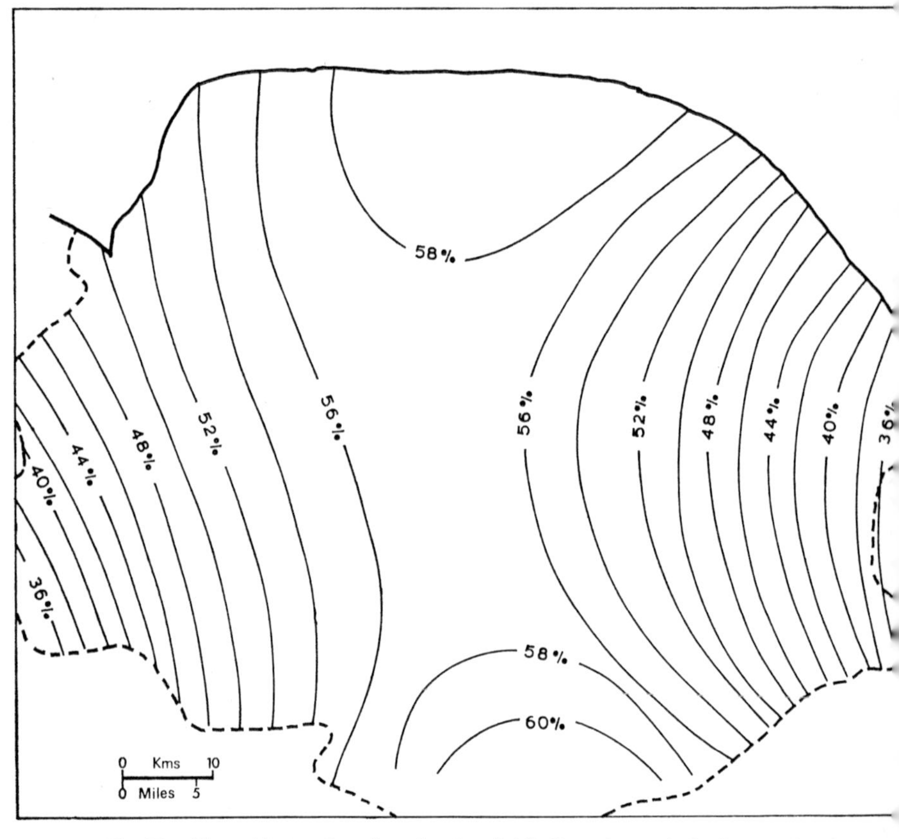

Fig 30. The cubic trend surface for the distribution of cereals in the county of Norfolk (based on parish data published by the Ministry of Agriculture, Fisheries and Food, 1968).

is a useful technique not only because it provides an economical description of the data but also as a hypothesis testing device, although this latter has been more often found in geological applications.[46] Like most techniques it is not free from problems.[47] One of the most intractable difficulties arises at the boundaries of the area. In fitting a surface to a number of control points it is assumed that the pattern shown by these points continues outwards in all directions. A simple linear surface, for example, will continue to have the same dip and strike defined within the area of the control points. For this reason there is no justification for interpreting the surface outside the area of these control points. These control points should also be evenly spaced. They do not have to occur at the intersections of a regular grid but should not be clustered in particular parts of the area and sparsely distributed elsewhere. An uneven scatter of points leads to undue emphasis being placed on the closely spaced points at the expense of the widely spaced points. Finally, it is difficult to assess the statistical significance of the surfaces.[48] The percentage of the total variance explained can be compared with that generated by fitting a surface of the same order to random data, but this requires a large number of iterations. The usual 'F' test that is used to compare the two variances—in this case the explained and the unexplained—is not appropriate as one of the parameters of the test is the number of degrees of freedom. We have no idea how many degrees of freedom are applicable in this sort of case. The degrees of freedom are the number of independently derived data values. In the example quoted it is clear that, to a greater or lesser extent, the value at one point is dependent on the values at neighbouring points: it is this assumption that allows us to find a surface to fit the data in the first place. Such dependence is known as spatial autocorrelation and when it occurs we have no measure of the number of degrees of freedom and we can not apply the 'F' test to this problem.

Despite the particular difficulties, there are clear advantages in using some type of contouring or surface-fitting technique with agricultural data. These techniques emphasise the continuity of change that is evident in many aspects of agriculture. The use of such a continuous surface to describe agriculture, rather than fitting the data into a number of discrete regions, will often be less misleading and may lead us to generate more accurate and realistic hypotheses to explain these patterns.

CHAPTER SIX

Explanation and Inter-relationships in Agricultural Geography

THE previous two chapters, concerning regionalisation methods and some alternatives to regionalisation, have shown how much effort in agricultural geography has been devoted to more efficient description of patterns of agriculture. In this chapter we will concentrate upon attempts to go beyond the task of ordering and classifying data, to one of seeking explanations.[1] Harvey illustrates two lines of enquiry which are used to seek explanations in geography and many other disciplines: enquiry by deductive and by inductive reasoning. In our discussion of theories of agricultural location the deductive approach is paramount: a theory is established and hypotheses are then formulated and tested using this theory as the framework. Theory 'provides the sieve through which myriads of facts are sorted and without it the facts remain a meaningless jumble'.[2] The importance of theory as a framework for the collection and sorting of facts is echoed by Berry[3] when examining the philosophy of economic geography. Despite its logical attractions, deductive reasoning is not characteristic of geographical explanation. Whether the lack of theory in geography[4] is the cause or the effect of the popularity of alternative approaches to geographical problems need not concern us here, but, measured in terms of weight of geographical publication, reasoning within the constraints of formulated theories remains insignificant.

Far more common in geographical writing is the approach recommended by Hartshorne[5] and echoed by Lukerman, who states: 'research in geography should begin with the description of geographic phenomena . . .'[6] This begins with the identification of a general field of study and often the implicit establishment of a general hypothesis. Such a field of study might be an hypothesis postulating a relationship between some aspect of the physical landscape and agriculture. For example, Robinson *et al.*, in seeking to account for variations in the rural farm population density over the Great Plains of the United States, recommend the use of statistical correlation techniques.

> One may properly employ these statistical-cartographic techniques after he has established tentative descriptive hypotheses regarding the mutuality that may exist among distributions in an area . . .[7]

The next and most important stage is the collection of data and the generalisation and classification of these, often resulting in the delimitation of regions. In order to test the hypothesis or hypotheses the generalised data have to be compared, often by no more subtle means than visual map comparison. On the basis of this comparison the hypothesis is accepted or rejected. In an examination of the importance of labour inputs to farming in controlling the level of outputs, Manley and Olmstead[8] compare visually maps of farming scale in terms of value produced and net operating farm income with maps of labour input in terms of full-time operator equivalents. It is in this field of hypothesis testing that statistical techniques, particularly regression analysis, have achieved considerable importance in geography.

The two methods of reasoning may not be so far apart as they seem. Deductive methods are undoubtedly more respectable as they rely on the prior formulation of a theory which can then be tested by controlled experiment. Data collection and subsequent testing are confined to the field specified by the theory. Inductive methods of reasoning may not be without a theoretical structure, although it is not often explicit. The facts themselves may be difficult to identify without some pre-existing theoretical structure.[9] Bunge[10] points out that there is an infinite variety of facts around us and that any description of them must be highly selective. This selection may be made at random but is more likely to be made on the basis of what appears significant. Without some conceptual base we do not know what we are looking for when collecting geographical facts. Classification used to structure and simplify data after collection also implies preconceptions about how the data should be ordered and these preconceptions go towards making a theoretical base to the investigation. The objective of any classification, and even the method chosen to collect the data, help to specify the method of classification which is used.

The most obvious approach to the seeking of explanations in geography is to establish causes for particular effects or facts. We describe the agriculture of a particular area and perhaps classify the farms according to the major agricultural enterprise, drawing up regional boundaries or possibly trend surfaces of the resulting data. We then proceed to examination of causes of the distributions illustrated. The simplest, but now discredited, method[11] is to seek for deterministic physical explanations for all geographic distributions. But, as Jones has explained, the fact that determinism is no longer accepted does not mean that geographers can no longer look for cause-and-effect relationships. 'The laws we invent are generalisations covering a vast number of data, not the laws of command which strict determinism suggests.'[12] Thus we do not expect a particular cause to give rise to the same effect on every occasion but rather that there should be a given probability of an event following from a specified set of circumstances. It is the general lack of proven deterministic relationships in geography which has encouraged the adoption of statistical techniques based on probability. On the other hand there are obvious dangers in assuming cause and effect between any two variables. In a relatively simple bi-variate situation it is more likely that there is a *relationship* between the two variables, which vary together in a particular manner, without any evidence that the variations in one have actually caused the change in the other. The use of correlation techniques to establish the relationship is appropriate here. A correlation between soils and agricultural yield does not imply that soil variations have actually caused the spatial pattern of agriculture. Hidore[13] presents the hypothesis that the growing of cash grains in the mid-west of the United States is associated with level land. The known importance of mechanisation in the production of grain crops supports this hypothesis. On the basis of correlations between the percentage of flat land and the percentage of farms which are classified as cash grain farms in a number of sizes of area, it is possible to conclude that the pattern of cash grain farming is indeed spatially associated with the flatness of the land. In this case, through the medium of mechanisation, there is a strong suggestion of cause and effect but the use of the correlation technique cannot be used to support this.

The correlation coefficient* can vary between +1·0 in the case of perfect positive correlation, where an upward increment in one variable is matched by a similar upward movement in another variable, to −1·0, or perfect negative correlation, where an upward movement in one variable is equivalent to a downward movement in another. The various relationships can be illustrated by scatter diagrams where individuals are measured according to two characteristics x and y and the results plotted on graphs (Figure 31).

Using data recording the agriculture of Sweden by county for 1970, the scatter diagram (Figure 31A) shows little or no positive correlation between yield of two crops and agricultural population density. The crops of barley and hay are used because they are the only two grown in all Swedish counties and the lack of positive correlation is surprising in the light of Ricardo's thesis that as yields increase so will rural population density. The negative correlation between the percentage of land in the eight Swedish agricultural areas devoted to arable crops and to forestry is not unexpected (Figure 31B). If a major portion of an area is devoted to one land use, only a little can be left for any other.

* The correlation coefficient (r) is calculated by

$$\frac{\frac{\Sigma xy}{n} - \bar{x}\cdot\bar{y}}{\sigma_x \cdot \sigma_y}$$

The numerator is a measure of the covariance of x and y and by dividing by the product of the standard deviations of the distributions of x and y the measure of covariance is made independent of the units of measurement.[14] If the values of x and y can only be ranked, then the alternative of the Spearman's Rank correlation coefficient can be used.[15] This statistic also has values ranging from +1·0 through 0 to −1·0.

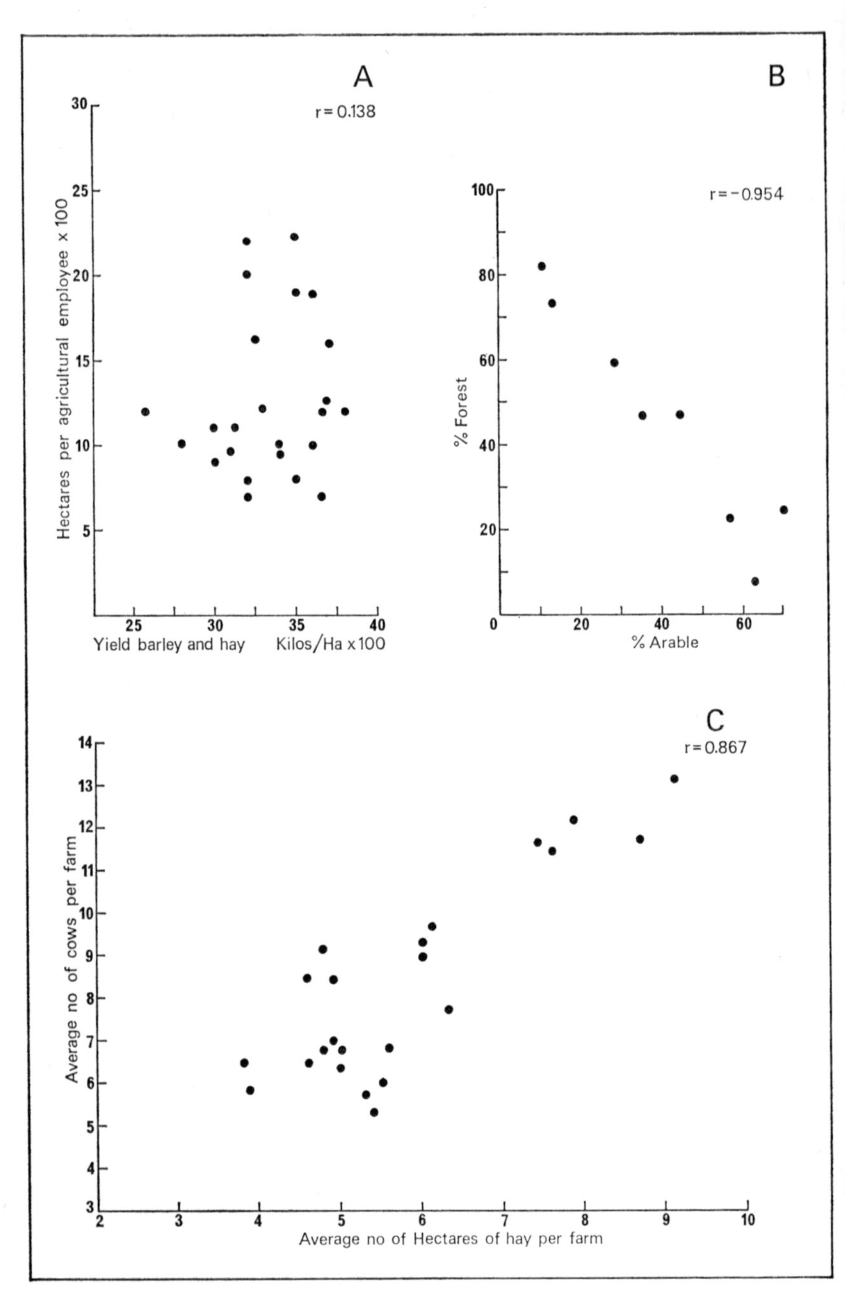

Fig 31. Scatter diagrams showing three types of inter-relationship between variables. In the case A no clear relationship is evident and a correlation coefficient of +0·138 results. The relationship between area of arable and forestry land (B) is much closer, but negative, while a close but positive relationship is found between the number of cattle and the area of hay produced (C). (All data from the *Swedish Statistical Yearbook for Agriculture,* 1971.)

Such closed data systems are common in agricultural studies and require care in treatment. Equally obviously the third scatter shows a fairly strong positive relationship between the average amount of land per farm devoted to a hay crop and the average number of cows on that farm (Figure 31C). There is no suggestion that the number of cows is in any way controlled by the acreage of hay or vice versa. It may be that large farms have a lot of both variables and small farms only a little of each. This scatter diagram simply suggests that the two variables vary together in a particular manner.

With a large number of variables a correlation matrix can be produced showing the correlation values between each variable and all others. From this matrix a linkage structure can be shown, linking each variable to the variable with which it is most closely correlated. From agricultural statistics for the 509 parishes in the county of Norfolk for which the Ministry of Agriculture compiled information in 1968, the correlation between the labour inputs (standard man-days) for all agricultural enterprises produces such a linkage structure (Figure 32). The direction of the arrow indicates the direction of the linkage; for example, barley is most closely correlated with sugar beet, and sugar beet with barley. From this linkage analysis groups of correlated enterprises can be identified. If, instead of linking variables together in terms of the extent to which they covary in a number of parishes, we had linked parishes together in terms of their inter-relationships measured over a number of variables, we could use this type of linkage analysis for purposes of classification.[16] This linkage matrix uses only positive correlations: the strengths of the links vary considerably and in some cases the correlations are very small. For example, with the exception of a very small positive correlation with rough grazing, all the correlations including vegetables are negative, suggesting that the distribution of vegetables is unlike that of any other enterprise listed. On the other hand barley, wheat, sugar beet, stock-feeding crops and dairy cattle make up a strongly knit group which tend to vary together throughout the county. Despite the fact that pigs and poultry are usually linked in discussion of agricultural geography, in terms of the spatial distribution of labour inputs to these activities, they have very little in common apart from a limited requirement for land as they do not depend on grazing. If we regard the 500 or so parishes of Norfolk as a sample of all possible parishes it is

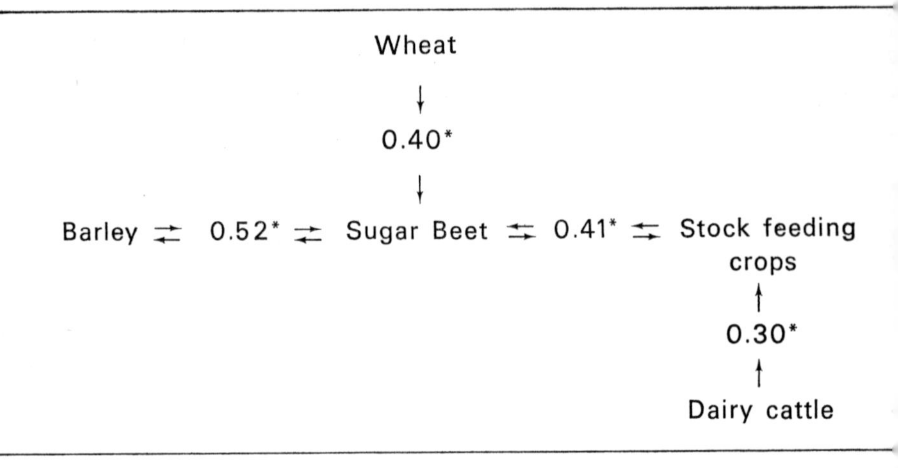

Poultry → 0.005 → Pigs → 0.21* → Grass ⇄ 0.73* ⇄ Beef cattle

Vegetables → 0.02 → Rough grazing ⇄ 0.70* ⇄ Sheep

↑ 0.27* ↑

Other crops

Potatoes ⇄ 0.32* ⇄ Peas

Orchards ⇄ 0.44* ⇄ Small fruit

↑ 0.11 ↑

Horticulture

Fig 32. Correlation linkage of Norfolk agricultural variables, 1968. Each variable is linked to the variable with which it is most closely positively correlated. The direction and the strength of the correlation are indicated by arrows. Correlation coefficients which are statistically significant at the 95 per cent level are indicated by asterisks.

then appropriate to apply a significant test* to such correlation coefficients. Correlations which were found significant at the 95 per cent confidence level are indicated by an asterisk on the linkage (Figure 32).

REGRESSION TECHNIQUES

As a technique for testing hypotheses, regression analysis goes substantially further than correlation. In linear regression we specify two variables, one the dependent and the other the independent variable. In the hypothesis which we wish to test one variable is set up as being dependent upon another: not only do they vary together but they do so because one causes variations in the other.† Unless it is possible, or desirable, to specify a

* The significance of a correlation coefficient is found by:

$$t = \frac{r \cdot \sqrt{(n-2)}}{\sqrt{(1-r^2)}}$$

where r = the correlation coefficient
n = the number in the sample.

The t statistic can be looked up in standard tables (Lindley & Milner, 1964) with the appropriate sample size minus two to give the size that the correlation coefficient has to reach in order to be statistically significant at a given confidence level. The 95 per cent confidence level implies that a correlation coefficient of this size is likely to occur by chance only five times in every hundred.

† In linear bivariate regression we expect the relationship between the dependent variable (y) and the independent variable (x) to take the general form of the linear function:

$$y = a + bx$$

where a is a constant and can be derived from:

$$a = \bar{y} - b\bar{x}$$

This value gives the position of the intercept of the regression line on to the y axis.

$$b = \frac{N\Sigma xy - (\Sigma x)(\Sigma y)}{N\Sigma x^2 - (\Sigma x)^2}$$

gives the regression coefficient or the slope of the regression line in the form of the tangent of the angle between the x axis and the regression line.

When dealing with sample values both a and b are estimates of the population values and are subject to the usual sampling errors. The regression line is therefore surrounded by confidence bands within which the regression line for the population would lie. The position of these confidence bands is decided by the size of the sampling errors of a and b. A full analysis of regression techniques can be found in King.[17]

dependent and an independent variable, regression should be replaced by correlation.

Suppose we wish to test the hypothesis that the level of mechanisation of farming is dependent upon the amount of arable land. Again using 1966 data for Swedish agricultural regions, the number of tractors per holding is plotted with the percentage of all land devoted to arable crops (Figure 33). The regression equation provides a best-fit straight line through the data points showing the extent to which the number of tractors is 'controlled by' the acreage of arable land. The total variance of the y variable is provided by a measure of the scatter of points around the mean of y or $\Sigma(y-\bar{y})^2$. The variance around the regression line is provided by comparing each y value to its estimated value on the regression line for a given x value. Thus $\Sigma(y-y^e)^2$ is the residual variance after fitting the regression line. The variance explained, or accounted for, by the regression line is $\Sigma(y-\bar{y})^2 - \Sigma(y-y^e)^2$. The variance explained as a percentage of the total variance gives a measure of the success of the regression hypothesis. It can also be derived from the square of the correlation coefficient—in this case $0{\cdot}94^2 = 0{\cdot}8872$ indicating that 88·7 per cent of the variance of the use of tractors can be accounted for by variation in the distribution of arable land.

Most situations, and agriculture is no exception, are multivariate in nature with many variables all interacting with, and to some extent causing changes in, each other. Agricultural patterns are a consequence not only of soil, but also of climate, markets, age and experience of the farmers and many other variables which may or may not be quantifiable and susceptible to statistical analysis.

MULTIPLE REGRESSION

In the previous example of simple linear regression the hypothesis was that variations in the dependent variable could be accounted for simply by variations in one independent variable. Multiple regression techniques permit inclusion of a number of independent variables and the regression equation takes a more complex form allowing the dependent variable to be a function of a number of other variables, for example: $y = a+bx+cz$. In this case the independent variables x and z are regressed with the dependent variable. A best-fit plane is fitted to the scatter of points in three dimensions. With more than two independent variables

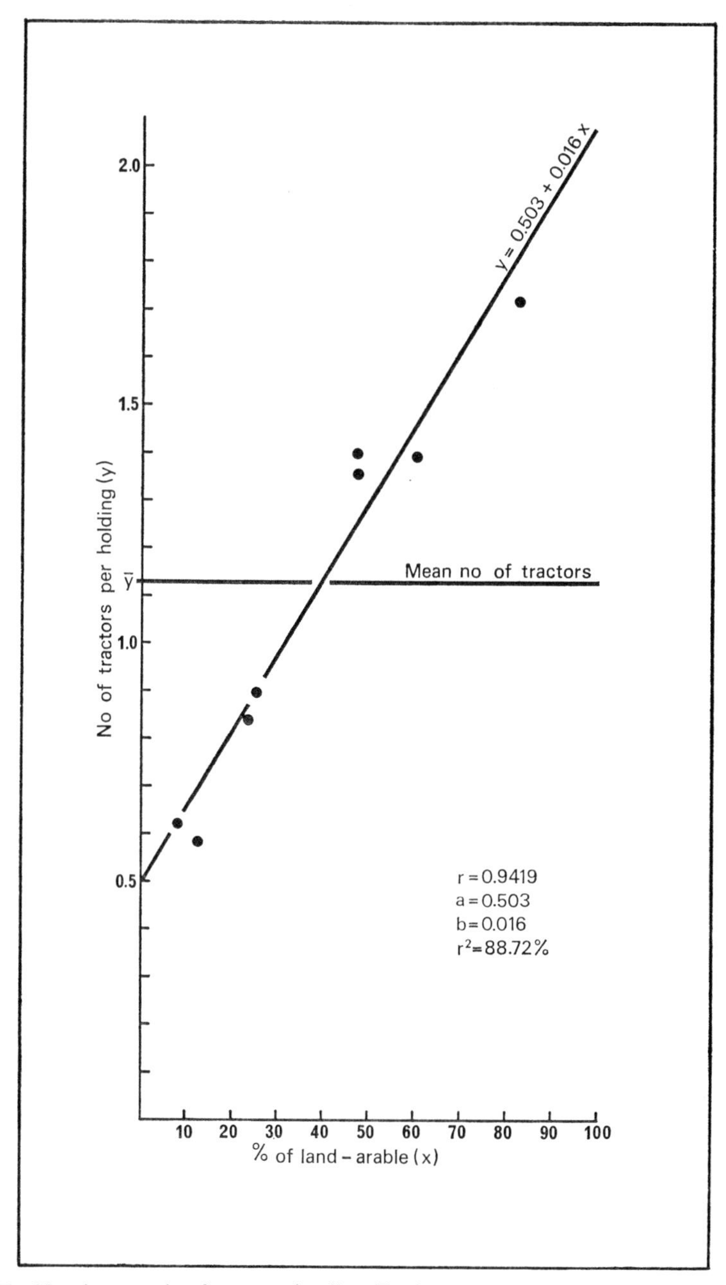

Fig 33. An example of a regression line. The hypothesis is that the percentage of arable land in an area has some control over the number of tractors in that area. (From data published in the *Swedish Statistical Yearbook for Agriculture*, 1971.)

such geometric representation is not possible but the same principles apply in 'n' dimensional space.

Robinson *et al.*[18] in an investigation of the rural farm population density in a part of the Mid-West of the United States hypothesise that the level of annual precipitation, distance to main urban centres and the percentage cropland can be taken as determinants of the variations of rural farm population density. The multiple regression equation in this case was found to be:

$$x_1 = -4{\cdot}854 + 0{\cdot}304x_2 - 0{\cdot}012x_3 + 0{\cdot}080x_4$$

where: x_1 = the rural farm population density,
x_2 = the level of annual precipitation,
x_3 = the distance to urban centres over 10,000 population,
x_4 = the percentage of total area in cropland.

R^2 or the explanation of the regression model was $+0{\cdot}81$ which means that 81 per cent of the variance of x_1 was explained by variation in x_2, x_3 and x_4.

PRINCIPAL COMPONENTS ANALYSIS

Agriculture includes a very large number of variables of greater or lesser interest. When examining the causes of variation it is useful to reduce the number of variables, not by eliminating redundant or insignificant ones as in multiple regression, but by replacing the whole matrix of variables by a matrix made up of a more limited number of new variables which will stand in place of the original ones. What is needed is a type of classification or generalisation of the variables so that we can reduce the magnitude of the variable matrix to a level which we can comprehend.

This can be achieved by applying one of many factor analytic solutions, the simplest of which is known as principal components analysis. The starting point of this analysis is the correlation matrix of the variables. Taking five variables we can draw up such a matrix (Table 17), although this particular example is simplified for reasons which will become clear later. If we represent one of these variables, sugar beet, as a vector or line of unit length from an origin (o), all other variables can be represented as similar unit vectors from the same origin arranged so that the angle between any two vectors is equivalent to the strength of the correlation

TABLE 17 Correlation matrix of five agricultural variables

	Wheat	*Sugar beet*	*Barley*	*Pigs*	*Potatoes*
Wheat	1·0				
Sugar beet	0·906	1·0			
Barley	0·766	0·966	1·0		
Pigs	−0·087	0·342	0·574	1·0	
Potatoes	−0·50	−0·819	−0·940	−0·819	1·0

coefficient between them. If, for example, a line is projected vertically from the end of the sugar beet vector on to that for wheat a right-angled triangle is formed and the two vectors intersect at an angle of α (Figure 34). The cosine of α is adjacent/hypotenuse = $<1/1$ which is some value between 1·0 and 0. The greater the angle the smaller the value of the cosine until if α equals ninety degrees, the cosine equals 0. The cosine can be readily interpreted as the correlation coefficient between the two variables. In our example the correlation between wheat and sugar beet is +0·91 and the corresponding cosine shows that the angle between the vector for wheat and the vector for sugar beet is 25°. If we include all the variables in our matrix (Table 17), the pattern

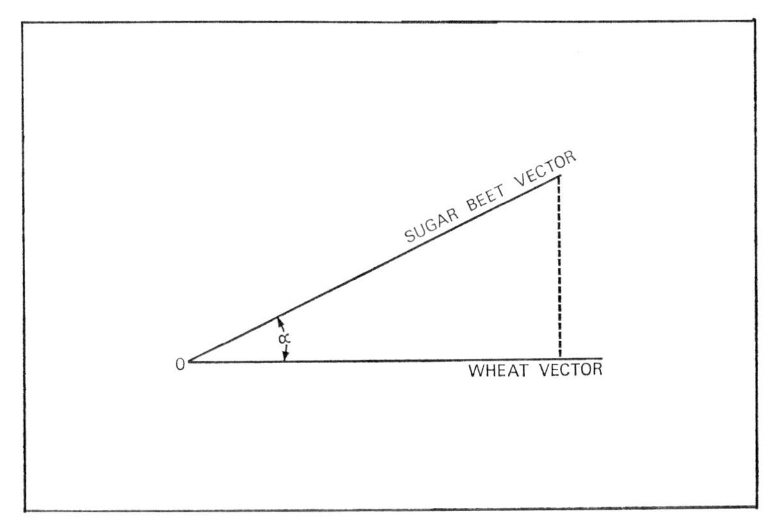

Fig 34. The relationship between two variable vectors with the angle between them, α, representing the correlation coefficient between the two variables.

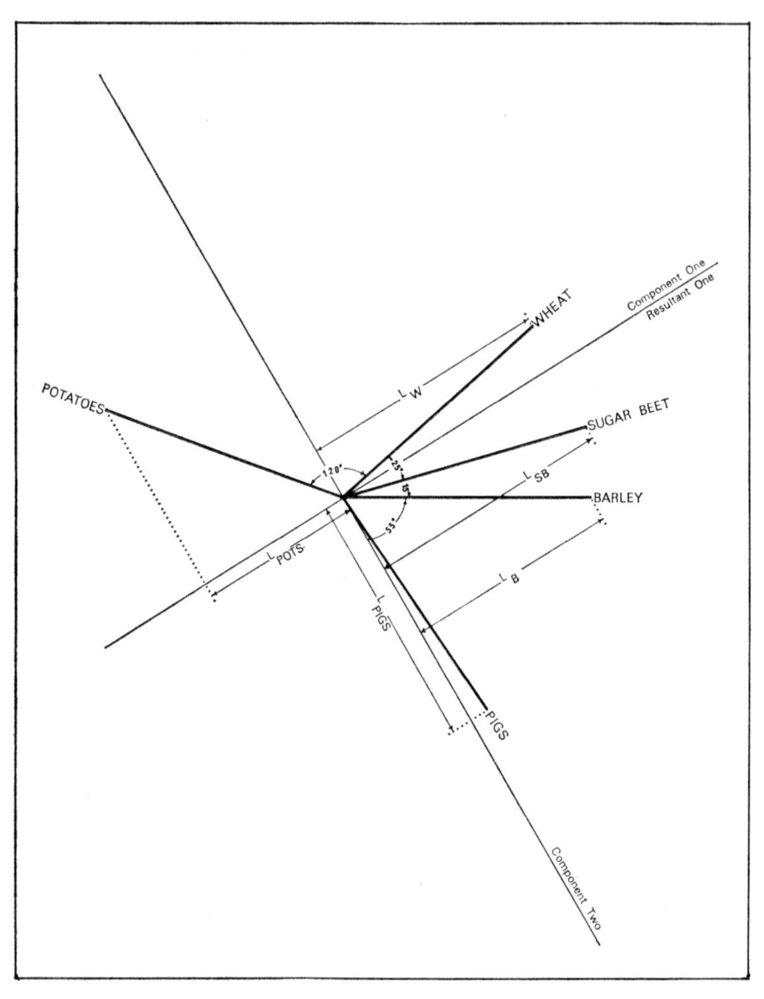

Fig 35. The relationship between five agricultural variables and the first two principal components. The variable vectors are shown only on one side of the origin for clarity. The lengths, L_w, L_{sb}, L_b, L_{pigs}, and L_{pots} represent the loading of the variables on the components.

of vectors is shown by Figure 35. It is now clear why the values of the correlation matrix had to be chosen with care to enable all the angles between the vectors to be represented on a plane. In reality the correlation between five such variables would probably require the use of an imaginary five-dimensional diagram if all the vector angles were to be correct.

In order to resolve the inter-relationships between the vectors we need to relate them to a single reference line or vector. All variables can be identified according to their angular distance from this reference vector which is known as a common factor vector. Child[19] illustrates this by the example of the spokes (or vectors) of an umbrella which are provided with a common factor vector in the handle. The individual variable vectors can be related to the factor vector by the same method as the variables are related to each other; the cosine of the angle between them is equivalent to the correlation coefficient between them. Through the cluster of variable vectors the factor vector is placed so that it is in the position of best-fit through the vectors. The sum of the lengths L_w, L_{sb}, L_b, L_{pigs}, L_{pots}, must be maximised. This common factor vector is known as the first resultant. The cosines of the angles between the variable vectors and the factor vector are called the factor loadings. These are listed in Table 18. The loading tells us how the factor vector is made up, how like each of the original five variables it is. In order to avoid confusion with other techniques we will call this first resultant the first principal component or first component. Because of our simplified example, all five variables can be expressed in two dimensions and can be replaced by two components, one of which we have already defined. The second component is erected at right angles to the first, passing through the origin, and a second set of loadings are derived for this new component (Table 18).

In a more realistic case, where the five variables do not lie in only two dimensions, the vectors would form a type of five-dimensional ellipse; the first component identifies the longest axis of this ellipse because of its requirement of maximising the summation of Ls. The second component identifies the next longest axis but is placed at right angles to the first, and so on until all the dimensions of the vector space are defined. In principal component analysis there are as many components as there are variables, although some of these are more or less redundant in terms of the amount of the total variation among the variables that

TABLE 18 Interpretation of principal components

Variable	*Component 1*		*Component 2*		Σ^2 *of component loadings**
	Angle	*Corr. coef. (loading)*	*Angle*	*Corr. coef. (loading)*	
1. Wheat	10°	+0·98	80°	+0·17	1
2. Sugar beet	15°	+0·97	106°	−0·24	1
3. Barley	31°	+0·86	121°	−0·52	1
4. Pigs	85°	+0·09	175°	−0·97	1
5. Potatoes	130°	−0·64	40°	+0·77	1
Sum of squared loadings (Eigenvalues)		3·06		1·89	5

*The summed squared component loadings on the two components for each variable should equal one and the summation of all the component loadings equals five, or the number of variables involved. Small rounding errors are involved in this table.

they account for. A more mathematical treatment of principal components analysis will be found in King.[20]

Components are of two types: common components with high loadings on most or all of the variables and unique components with loadings which are low on most of the variables and high on a limited number or even only one, suggesting that this variable is orthogonal to or uncorrelated with all the others and that a separate component vector is required to account for it. As the objective of all factor analytic methods is to reduce the number of variables, or dimensions, in the problem and to examine composite, average components replacing the original variables, we are primarily interested in the common components rather than in the unique ones.

To illustrate an actual application of principal components analysis, especially the extraction of the first, general, component, we will use the same data as in the correlation linkage analysis: all the variables except the last are expressed as the percentage of the total standard man-days in each parish devoted to each agricultural type. The last variable is the number of standard man-days per acre of agricultural land (crops and grass), which therefore provides a measure of the degree of labour intensity within the agriculture of the parish. Before pursuing the components analysis of these data further, what assumptions are we making about the nature of the data, and are these assumptions reasonable? Non-normal data are acceptable for this type of analysis, provided they are capable of transformation to normality.[21] As the data which we are using are percentages of the total labour inputs devoted to each agricultural enterprise within the parishes of Norfolk, it is probable that some will have skewed distributions, especially those enterprises which have a local occurrence within the county. The exclusion of other crops, rough grazing, sheep, pigs and poultry was necessary because they have particularly skewed distributions and also because the correlation matrix showed them to be poorly correlated with the remainder of the data. The next requirement concerns the nature of the inter-relationship between the variables. It is clear from the diagram of the variable vectors (Figure 35) that the relationship between the variables has to be linear; serious curvilinearity amongst the inter-relationships makes this form of component analysis impossible.

Finally there is the question of the range of values represented for each of the variables. In order to reduce all the variables to the same scaling, the scores of each parish on each variable are standardised by dividing the difference between each score and the average of all scores by the standard deviation of the distribution, thereby expressing each score in standard deviation units away from the average.* The mean of this transformed distribution is zero and the standard deviation unity. Despite the considerable contrasts among the variables these have now similar scaling.

After conducting the component analysis, for which there are numerous computer programs available,[22] the components are extracted as in Table 19; there are as many components as there are variables. The eigenvalues, or variance extracted by each component, can be calculated by squaring and summing the loadings on that component (Table 18). As the total variance is fifteen, or the total number of components extracted, the eigenvalues can be used to give us an indication of the relative impor-

TABLE 19 Components and eigenvalues of principal component analysis

Component number	*Eigenvalue (component variance)*	*% of total variance*	*Accumulated % of total variance*
1	3·965	26·43	26·43
2	1·977	13·19	39·62
3	1·612	10·75	50·37
4	1·255	8·36	58·73
5	1·071	7·17	65·90
6	0·957	6·38	72·28
7	0·858	5·72	78·00
8	0·658	4·38	82·38
9	0·641	4·28	86·66
10	0·523	3·49	90·15
11	0·472	3·15	93·30
12	0·342	2·28	95·58
13	0·273	1·82	97·40
14	0·229	1·52	98·92
15	0·161	1·08	100·00
Total variance	15	100	

(From data published by Ministry of Agriculture, 1968)

* The standard score $= \frac{x - \bar{x}}{\sigma_x}$

In order for this standard score to have any meaning the distribution of the variable being transformed must be normal so that the standard deviation has some meaning.

tance of each component in accounting for this variance. The first component extracts: $3{\cdot}965/15 \times 100 = 26{\cdot}43$ per cent of the total variance.

Two problems now arise: first, what do these components mean in terms of the original matrix of variables, and secondly, how many of these components can be considered general and therefore useful within the objectives of component analysis? There are several possible answers to the second question because we do not expect the boundary between general and unique components to be abrupt; where the line is in fact drawn will depend to some extent on judgement. With a principal components solution to a problem with between twenty and fifty variables Kaiser's criterion is often applied, whereby those components with an eigenvalue of greater than one are considered significant. In this example we would take the first five components as being general and the remaining ten as being unique and therefore of little interest. An alternative is to use the scree test recommended by Cattell[23] which requires plotting the eigenvalues against component numbers and selecting as the cut-off the point where the slope of the plotted graph becomes linear (Figure 36). This test would give us seven general components. One important point not regarded by either test is that eigenvalues which are very similar should not be separated. If we consider these components as orthogonal axes through a multi-dimensional cluster of points, the eigenvalues are a measure of the extent of the clustering around the axes, or the variance explained by these axes. Taking a rugby or an American football as an example, the first, major axis of the cluster is obvious and easily identifiable. The second, placed at right angles to the first, could be any diameter through the circular mid-section of the ball, its positioning in the case of a perfectly circular section being arbitrary. Two or more factors with very similar eigenvalues suggest that the placing of these axes in the multi-dimensional space is arbitrary and the scatter of points around each of the resulting components is almost identical. As we shall see, this makes for difficult problems of interpretation when giving names to these components. It would clearly be wrong to draw a distinction between general and unique components in the middle of such a group of similar eigenvalues.

The question of the interpretation of the components is very much more difficult and is an exercise which should be approached with great caution. If the objective of the component analysis is to produce a limited set of new variables in order to carry out some sort of classification or regionalisation exercise on the individual cases,[24] there is no need to interpret the components themselves as they have been based on the variables originally used and can be taken as a parsimonious version of this original

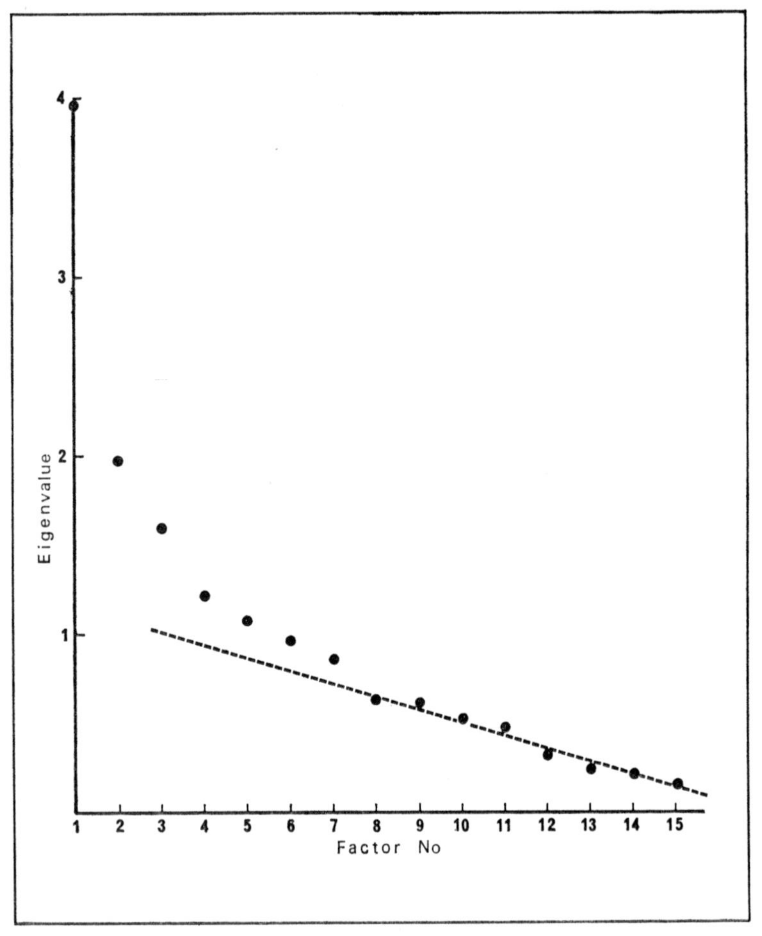

Fig 36. A diagrammatic representation of the operation of the scree test on Norfolk agricultural variables, 1968. After a certain level the eigenvalues fall approximately on a straight line.

variable matrix. If we are seeking explanations of particular patterns of agriculture by the identification of underlying factors in its distribution, this interpretation of components may be more important. The difficulty in interpretation can be visualised by referring back to Figure 35. In placing the first component so that it is the best-fit through all the variable vectors, all these vectors have had some influence on the orientation of the component and, apart from the unlikely event of a variable vector lying at right angles to the component vector, all have some correlation with this component. In the case of general components this is exactly what one would expect and indeed require, but as a result the component cannot be related simply to one or to a limited number of variables and it is therefore difficult to interpret it.

As component loadings are correlation coefficients they can be subjected to a significance test. With a sample size of 500 a correlation coefficient has to reach at least ±0·115 to be significant at the 99 per cent confidence level.* With factors other than the first, correlation coefficients may not be high enough due to increased intrusion of unique variance, or variance within the variables themselves rather than from the inter-relationship between them. As a result Burt and Banks[25] provide a more stringent test which takes into account the number of variables used in the analysis and the number of the components being tested.

$$SE_L = SE_{cc}\left(\sqrt{\frac{n}{n+1-r}}\right)$$

where: SE_L = standard error of loading,
SE_{cc} = standard error of correlation coefficient,
n = number of variables,
r = the factor number.

Vernon[26] suggests that double the standard error of the loadings can be used to identify truly general components. Only those components which have at least half the variables with a loading of greater than twice the standard error can be considered general. The standard errors of the loadings with 15 variables and 500 individuals are indicated in Table 20.

In our example the first four components can be considered general (Table 21), while the remainder are unique. Component

* See footnote, page 177.

TABLE 20 Standard errors of component loadings with 15 variables and 500 individuals

Factor number	*Standard error of loading*	*Twice standard error of loading*
1	0·115	0·230
2	0·119	0·238
3	0·123	0·246
4	0·128	0·256
5	0·134	0·268
6	0·141	0·282
7	0·148	0·296
8	0·157	0·314
9	0·168	0·336
10	0·182	0·364
11	0·199	0·398
12	0·223	0·446
13	0·257	0·514
14	0·315	0·630
15	0·445	0·890

Standard error of correlation coefficient = ±0·115

one is a truly general factor with all but one of the variables being significantly correlated with it. The nature of this general factor can be derived by examination of the strengths and directions of these loadings. The highest loading of all, –0·836, is with the variable measuring standard man-days per acre in the parish, and other high negative loadings are found with crops which have a high labour requirement, particularly horticulture and small fruit. The highest positive loadings are with the labour-extensive cereals, crops and grass. A possible exception to this pattern is sugar beet which has a high labour requirement at the thinning-out stage which is often still done by hand. This first component seems to indicate a major underlying component of variation in agriculture of Norfolk as being the requirement for labour or the intensity of labour use.

To assess the spatial importance of this new variable its distribution was mapped (Figure 37). The importance of each of the original variables in any parish is provided by the standard score of that variable in the parish. The extent to which this variable has contributed to the new component is provided by the component loading. We therefore multiply the standard score for each variable in the parish by the appropriate loading for the first

TABLE 21 Principal component analysis of agriculture in Norfolk

Variable	*Components*								
	1	*2*	*3*	*4*	*5*	*6*	*7*	*8*	*Commun-alities†*
heat	·400*	·595*	·231	·011	·104	·030	·102	·347*	·567
rley	·781*	·111	−·278*	·193	−·106	·145	−·137	−·085	·737
reals	·474*	·102	−·199	·273*	−·116	·125	·651*	−·312	·347
tatoes	−·364*	·589*	·311*	−·210	·177	·034	·049	·221	·620
ock feeding	·605*	−·152	−·032	·256*	·020	−·280	·271	·462*	·456
gar beet	·551*	·591*	−·156	−·006	·033	·228	−·234	−·198	·667
chards	−·366*	−·066	·387*	·473*	−·520*	·047	−·060	·016	·512
nall fruit	−·494*	·152	·444*	·412*	−·255	−·026	−·034	−·028	·634
getables	−·436*	−·094	−·414*	−·547*	−·398*	−·196	·168	·097	·669
as‡	−·211	·315*	·551*	−·258*	·196	−·314*	·230	−·341*	·514
orticulture	−·454*	−·259*	−·088	·303*	·570*	·390*	·233	·002	·373
ass	·526*	−·489*	·461*	−·237	−·096	·245	·089	·013	·784
iry cattle	·378*	−·405*	·169	·139	·342*	−·588*	−·318*	−·063	·355
ef cattle	·484*	−·467*	·457*	−·315*	−·040	·299*	·099	−·009	·760
ID/acre§	−·836*	−·242*	−·202	·079	·141	·036	·071	·037	·804
. of signifi-nt loadings	14	9	8	8	4	4	2	3	
	general components				unique components				

* Indicates a component loading significant at the 95 per cent level.

† The communalities (the sum of the squares of the component loadings) are calculated for the first four components only.

‡ Peas for vining.

§ Standard man-days per acre.

(From data published by the Ministry of Agriculture, 1968)

component and the summed result is known as the component score, or the score for this parish on the first component.*

The spatial distribution of this first, major, component confirms that it is a component concerned with the intensity of labour use in agriculture. In most of central Norfolk the scores generally indicate a low level of labour use associated with the major Norfolk land-use pattern of wheat, barley and sugar beet. The

*

$$cs_{jk} = \sum_{i=1}^{n} ss_{ji} \cdot cl_{ik}$$

where cs_{jk} is the component score for parish j for component k, n is the number of variables, ss_{ji} is the standard score in the parish j of variable i and cl_{ik} is component loading of variable i on component k.

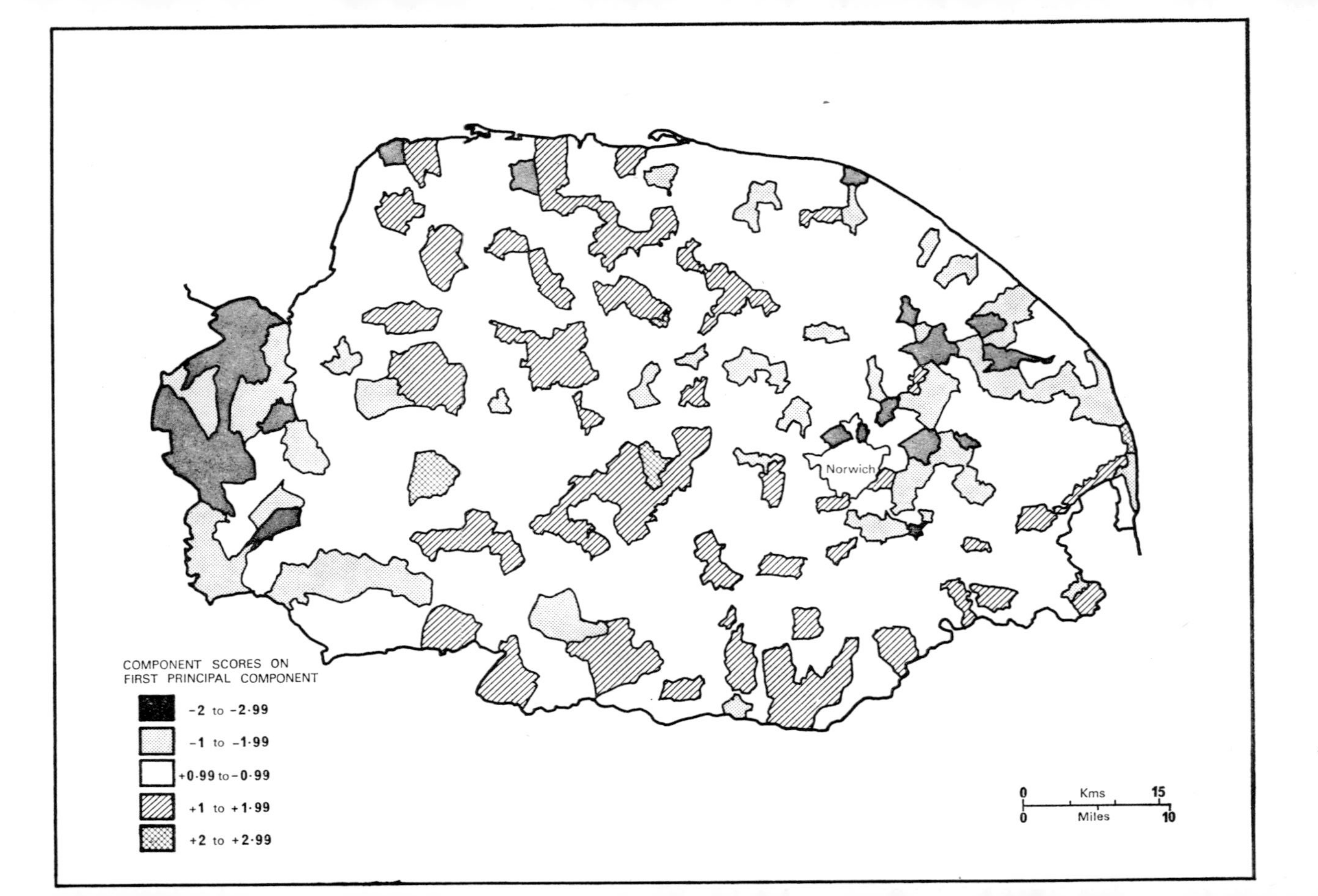
Norwich
COMPONENT SCORES ON
FIRST PRINCIPAL COMPONENT
−2 to −2·99
−1 to −1·99
+0·99 to −0·99
+1 to +1·99
+2 to +2·99
0 Kms 15
0 Miles 10

labour intensity level rises sharply in the Fens and in the vegetable and small fruit-growing areas in the north-east of the county.

There remains one difficulty with interpreting such a map. The component scores are derived by multiplying the standard score of each parish on each original variable by the loading of that variable on the component. A high-scoring parish may therefore achieve that score by either a very high standard score on a very insignificant variable or by a moderately high score on a significant variable. Since such different circumstances can produce the same score for a parish, component scores should be approached with caution.

From an examination of its loadings the second component seems to distinguish between sugar beet with potatoes and cattle but this and the remaining factors remain difficult to interpret.

FACTOR ANALYSIS

Principal component analysis does not require *a priori* hypotheses about the nature of the components being extracted nor does it distinguish between the variables in any way; in locating the component axes they are all equally weighted. Any variable has three sources of variance which together make up the total variance for which the components are attempting to account. The most important from the point of view of extracting general components is that type of variance which reflects the association of that variable with others, or the *communality*. In the example used the communality of wheat can be interpreted as that proportion of the total variation in the labour inputs to wheat throughout the county which can be associated with similar variations in other agricultural enterprises, while the unique variance is that which is peculiar to wheat alone. In the identification of general factors it is the strength of the communalities of all the variables which must be taken into account. The third source of variance is error variance within the variables, error in recording and calculating the labour inputs to wheat for each of the parishes. With little or no information about this error variance it is usually ignored.

Total variation within a variable can be expressed as: $V_{ti} = C_{vi} + U_{vi} + E_{vi}$, where C_{vi} is the communality variance in the ith variable, U_{vi} is the unique variance and E_{vi} is the error variance. In the identification of general factors the variables need not be considered equally important in helping to locate the component vectors. Referring to Figure 35, if the communality of wheat is low and the unique variance is high then wheat should not be as important in locating the first component as sugar beet if the communality of sugar beet as a variable is high and its unique variance low. In applying the best-fit criterion to the resultant vectors the variables may be weighted according to their communalities. Where the variables are not given equal weight the analysis is commonly called factor analysis. In order to allocate the strength of the communalities it may be possible to make some assumptions about the nature of the variables and the importance of the unique variance but this is difficult in agricultural problems. Another solution is to investigate the multiple correlation coefficient between one variable and all others. The higher this correlation coefficient, the greater the covariance between this variable and all others, and the greater the communality of that variable. Another alternative is to look again at the principal components solution. If we isolate the general components then the sum of the squares of the component loadings for each variable on these general components can be used as an indication of the communality of that variable. In our example the communalities of the fifteen variables are indicated in Table 21, taking the first four components as the general ones. The communalities are low for some crops, especially those which are unimportant within the agriculture of Norfolk (for example, other cereals with a communality of only 0·347 or 34 per cent of the total variance) and also for crops which have a fairly local distribution within the county such as horticulture (0·373). Taking these communalities as weights the factor analysis can proceed to fit four or more new factors through the variable space according to these weights, and the communalities are redefined for these new general factors. Because the original principal components have now been rotated to new positions aligned to smaller groups of variables, the new factors are often easier to interpret than the components. The factor loadings on the four factors (Table 22) indicate that the position of the first component has not altered significantly but the second factor is more strongly loaded on sugar beet, wheat and barley and negatively loaded on a number of activities, especially grass and beef cattle. This second factor can now be identified as representing the type of agriculture typical of much of central Norfolk. The third

TABLE 22 Factor analysis of Norfolk agriculture

Variable	*Communality*	*Factor 1*	*Factor 2*	*Factor 3*	*Factor 4*	
Eigenvalues		3·481	1·471	1·216	0·705	Σ6·873
Wheat	·351	0·345	0·310	0·353	0·108	
Barley	·774	0·724	0·380	−0·264	−0·188	
Cereals	·187	0·360	0·192	−0·098	−0·105	
Potatoes	·507	−0·302	0·154	0·564	0·273	
Stock feeding	·281	0·489	0·042	−0·149	−0·135	
Sugar beet	·536	0·481	0·530	0·114	0·107	
Orchards	·373	−0·281	−0·161	0·230	−0·465	
Small fruit	·623	−0·415	−0·081	0·476	−0·465	
Vegetables	·247	−0·385	−0·039	−0·180	0·256	
Peas*	·280	−0·145	−0·050	0·472	0·182	
Horticulture	·270	−0·440	−0·172	−0·209	−0·054	
Grass	·829	0·631	−0·653	0·046	0·004	
Dairy cattle	·134	0·298	−0·156	−0·132	−0·060	
Beef cattle	·640	0·549	−0·577	0·015	0·075	
SMD/acre†	·845	−0·863	−0·195	−0·248	−0·008	
Total communality	6·877					

* Peas for vining.
† Standard man-days per acre.

(From data published by the Ministry of Agriculture, 1968)

factor is positively loaded on potatoes, vining peas, and small fruit, identifying crops of considerable, but rather local, importance within the county. The fourth factor, of relatively minor importance, remains difficult to interpret but appears to be making a distinction between those areas with vegetables and potatoes and those areas with orchards and small fruit, all four crops being labour-intensive. As a result of the establishment of communalities, and the weighting of the variables accordingly, the first three components of agricultural variation in the area have been more clearly established.

If we have further hypotheses about the number and nature of those factors we need to extract, we can conduct one of a number of more complex rotations on the factor vectors in order to locate them in association with particular groups of variables. There is no real reason why the factor vectors should be orthogonal, although if they are not so they should be used with caution for classification purposes as they will be no longer independent, uncorrelated variables. If there are clusters of variable vectors, and there is no reason why the centres of these clusters should be orthogonal, it is often more appropriate to locate the first reference vector or factor in the centre of the largest cluster and then others located according to other, less important clusters. This type of analysis is not often used in geography because it is usually not easy to make the necessary assumptions about the nature of geographical variables. The association between some of the many different techniques of factor analysis is described more fully in a book by Harman.[27]

An example of the application of rotated factor analysis is provided by Henshall,[28] although it is not clear how the communalities have been estimated or what has been the basis of the rotation used. In an attempt to explain variations in the pattern of agriculture in Barbados, the underlying structure of this variation was examined with the help of factor analysis in order to test the hypothesis that the basic cause of the variation was demographic factors within the island. Thirty-two variables were included in the analysis ranging from the percentage of the land under different crops, to slope and soil conditions and the importance of women as farmers. Twelve factors were extracted. Although there are obvious difficulties in interpretation of the factors for the reasons which we have already considered, the first three factors can be tentatively identified as urban influence, fragmentation of holdings and demographic situation (basically family size, number of children and the importance of migration within the family). This analysis provides a useful indication of the role of factor analysis as a hypothesis-testing device. In the context of complex causes of agricultural patterns such multivariate techniques have considerable advantages, although they require the assistance of computers. The dangers of their application to inadequate data and of glib interpretation and naming of factors must be appreciated.

In previous chapters there is an implied assumption that the analysis of land use is closely related to farm organisation. A study of 218 farms in southern England, the Midlands, East Anglia and the Welsh Borderland (by Munton and Norris[29]) is an attempt to test this assumption through a principal components analysis. The properties of the farms measured include a group concerned with land use, a group concerned with crop yields and fourteen variables dealing with the application of fertilisers. There is a further group of variables concerned with the size, structure and management of the farm and thirty-three variables concerning the physical characteristics of the farm, including soil, aspect, and drainage. It is not clear how some of these variables were measured and nothing is said about the nature of their distributions, some of which must be distinctly skewed. A total of eighty-two variables is used in all. A number of different analyses are conducted on these data using both R mode and Q mode analysis. The Q mode analysis enables the farms to be ordered according to components defined from the variables, and the R mode analysis allows the variables to be arranged according to components derived from the farms. The analysis identifies numerous inter-relationships among the physical properties of the land and the farm enterprises; for example, Figure 38 shows a sample of

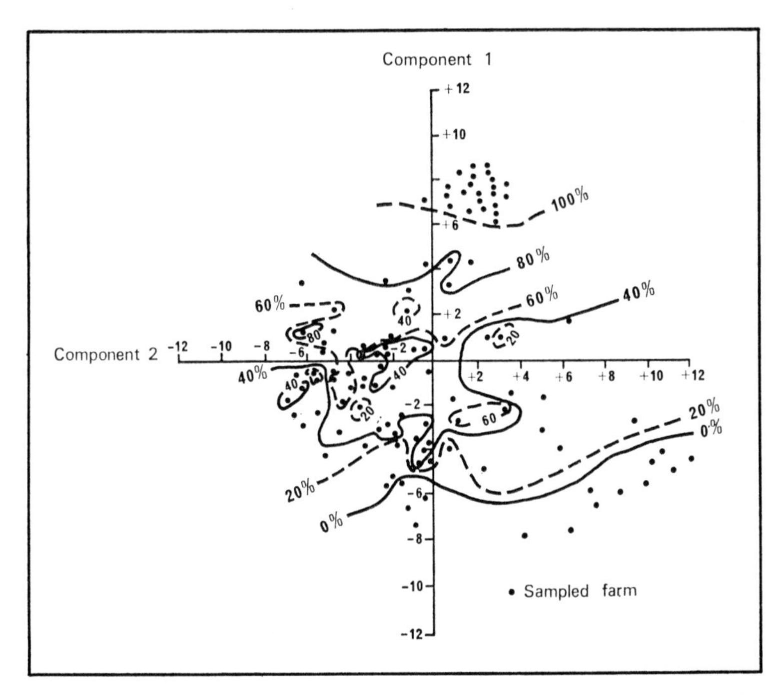

Fig 38. Farm percentage grass cover mapped on location of farms plotted according to their locations on the first two principal components. The isopleths indicate that component one shows a gradient from no grass to total grass.

farms plotted according to their scores on the first two components derived from only the land-use variables. The first component identifies a gradient from arable farming to pastoral farming with percentage grass cover having a loading of +0·86 and all arable enterprises having negative loadings. The percentage grass cover on the farms has been indicated by isolines on this diagram. The second component identifies a gradient from cereals to cash-root crops with variable loadings of: oats −0·63, barley −0·39, summer wheat −0·06, winter wheat +0·15, sugar beet +0·63, potatoes +0·65.

As we have seen, principal components analysis has considerable value for the generation of hypotheses. The elucidation of the major components of variance within an agricultural situation could suggest suitable hypotheses concerning the causes of such variation. Factor analysis, on the other hand, involving assumptions about the communalities and, if rotation is intended, considerably more about the inter-relationships between the variables, is best based upon existing hypotheses and can be used to test these. The existing concentration within agricultural geography on description rather than explanation is largely a result of the nature of the available data. Crop and stock data invite assumptions concerning their relationships to management and economic and physical factors without providing any basis for these assumptions. Studies like those of Munton and Norris[30] are made possible only by detailed farm surveys. Explanation therefore requires both multivariate data and multivariate methods of analysis of these data. Even with both available it is important that they are used to establish or test hypotheses rather than to supply yet more sophisticated descriptions. The example of agriculture in Norfolk, used throughout the latter part of this chapter, is little more than an efficient description of known data. The identification of the major component of variation as one reflecting variations in labour intensity throughout the county does little or nothing to *explain* this pattern, but the example has served to introduce the techniques which can be applied to more comprehensive data as, for example, by Henshall and Munton and Norris. Factor analysis and component analysis can only restructure the data as measured; the underlying components of variation have to be drawn from some combination of the variables included. It may, therefore, give us some idea of what to measure in future studies of agriculture. It is possible that the major element of variation of farms or areas may have been outside the frame of reference and the inclusion of such variance would completely relocate the component or factor axes. In this respect such analysis is very similar to classification procedures where, once a classification has been decided upon, classes result and, although they may be the most efficient such classes in a statistical sense, they have no fundamental validity outside that particular data set. Therefore, classification should never be used as an end in itself. Similarly, component and factor analysis can be used to examine the nature of the data variability and to suggest what are the important features of the particular environment being examined which should be of concern in future studies.

CHAPTER SEVEN

Agricultural Marketing

AGRICULTURAL marketing, encompassing the movement of goods from the farmer or producer to the processers, wholesalers, or directly to individual households, has no parallel in any other form of production. This uniqueness is evident in a number of ways. The first, and probably the most obvious, is the large number of producers and their great diversity in scale, specialisation and production methods. In addition there is great variety among the recipients of agricultural produce as it moves off the farm: producers sell to individual household units and to the large food-processing industries or to government marketing agencies. One fundamental characteristic of agriculture is that the producers outnumber the consumers—not the ultimate consumers, normally the total population, but rather the selling outlets available to, or utilised by, the farmer. In addition to these numerical characteristics, agricultural marketing also has special economic features. In von Thünen's isolated state, demand is fixed and predictable, and so, therefore, is supply: in the perfect world of such a model, supply can be matched to demand because a uniform and stable physical environment and the operation of farmers in a state of perfect knowledge are assumed. As supply is always perfectly matched to demand it is clear that prices are also stable. While this type of model has its uses, particularly in illustrating how certain variables operate in isolation from the effects of others, the reality of agricultural marketing is different on almost all counts. Demand is relatively stable, at least in the short term, although long-term changes in food tastes may have substantial effects on the marketing of agricultural produce and many of these taste changes may be stimulated, or in turn may stimulate, changes in the technology of agricultural production. The recent explosive growth of the frozen food market is an example. Demand for many products is stable or slowly rising and their prices inelastic. This means that the price people will be prepared to pay, for example for potatoes, will not rise much in the event of a shortage, nor fall much in circumstances of overproduction. On the other side of the coin is a highly variable supply. This provides a further exceptional aspect of agricultural marketing. Variable supply is a consequence of all the factors which von Thünen held constant—physical factors, especially weather and its fluctuations, and human factors involving information, the consequence of a great number of producers operating within a market system in states of incomplete knowledge of their competitors and of the expected state of the market at the time the product is really for sale. The situation is aggravated further by the long production process involved in agriculture. The delay between planting and harvest or between livestock rearing and marketing is so long that producers cannot react to changed market circumstances so as to have any effect on the marketing of, or profit on, a season's production.

Unreliability of physical conditions, marketing delays and lack of information tend to augment each other to compound the difficulties of agricultural production and marketing. As an example, assume a bad year for the production of brussels sprouts. Poor weather conditions perhaps have meant that many producers have achieved only small yields. As a result prices are fairly high, as high as the relatively price-inelastic demand situation can allow. Returns are high to those farmers who have managed a good crop. As a result of good returns to some farmers, a large number of farmers may be attracted to grow a few hectares of brussels sprouts in the following year. If this second year coincides with good weather conditions and crop yields there will be a glut of brussels sprouts on the market, and returns per hectare will fall drastically. The price will be unable to fall enough to dispose of all production while still covering production and harvesting overheads. The result will be that a proportion of the crop remains unsaleable, even direct to the consumer on a 'pick your own' basis, and the producers will bear this loss. In an extreme case, the following third year so many farmers will change to other crops with a more certain market that producers may prove to have over-reacted, resulting in a market shortage in year three. Such a cycle may continue, its regularity upset by random physical fluctuations in growing conditions. This example concerns a crop generally grown in small acreages without a great deal of capital investment. When commodities requiring a substantial level of capital investment are involved, as in for example the establishment of a high-quality dairyherd, farmers are unable to react to changes in the market conditions brought about by yield fluctuations from year to year, therefore returns to individual farmers are highly erratic. A further circumstance makes agricultural marketing a difficult and unstable process. Some commodities, more especially milk and eggs, are in fairly constant demand throughout the year, while production is far from uniform and storage is difficult. If left to operate freely, the milk market would vary from conditions of glut in summer to shortage in winter and the price would vary as much as the demand allowed. Because of the nature of lactation it is not possible to correct this situation without finding an alternative outlet for surplus milk in the summer and carefully controlling the distribution of milk to the different markets throughout the year.

Long- and short-term storage of agricultural produce has always been difficult or, for some commodities, impossible. Improvements in storage capability brought about by refrigeration, freeze-drying, canning and other food processing, as well as by the introduction of chemical pest control and storage environment control, have made the greatest of contributions to the seasonal price stability of agricultural markets. However, seasonal surplus for many crops has been replaced in some countries by an annual production surplus. In such circumstances storage can help only if the price of the stored commodity rises in future years and the quality of the stored product does not deteriorate. On balance it may be found that storage of surplus production provides no solution and may contribute to an accumulating problem for the future. Thus stockpiled American wheat has often built up to such proportions that it has had to be either destroyed, sold at uneconomic prices, or used as a part of the American overseas aid programme.

In the matter of transport costs the marketing of agricultural produce differs substantially from the pattern in the simple isolated state of von Thünen. Transport rates are far from simple and may be linked with many other features of agricultural production. Transport of agricultural produce may be subsidised in a number of ways either by the marketing organisation or directly by the purchasers who may also be the suppliers of feedstuffs and fertilisers, for example grain merchants. The possibilities of a return load of fertiliser or grain means that merchandising of grain and supplying of feeds and fertilisers can be economically run together; in order to retain the custom of a farmer in both these fields of operation, a subsidised transport rate may be possible. Transport rates may also be subsidised in many other ways, particularly by using short hauls to subsidise long ones. These factors together serve to confuse the pattern of the distribution of goods, while the quantities produced and their selling price are confused by imperfections of the market system.

These several reasons for the uniqueness of agricultural marketing form a justification for a substantial and ever-increasing degree of control imposed on the agricultural market system both by private consortia, either producers or purchasers, and by government, on a local and national or international scale. The degree of this interference in the free play of the market is itself a further aspect of the special nature of agricultural marketing. The large number and variety of producers, the contrasting scales of the available outlets, the erratic supply and stable demand, government intervention and the great diversity of agricultural products from which producers have to choose, all make agricultural marketing of considerable interest. As the efficiency of the marketing process will have a substantial effect on the price of food, the marketing techniques adopted for the various commodities must also have a substantial effect on costs and standard of living.

THE NATURE OF THE MARKET

Many changes have occurred in the competitive structure of markets in which farmers sell their products.[1] Together these changes have drastically reduced the number of controlling interests in marketing of agricultural produce while the number of producers has declined relatively slowly. The effect of this on the producer may be more apparent than real, but at least farmers see themselves at a distinct disadvantage when faced with a large and economically powerful buyer of their produce.

We can identify three stages of agricultural marketing through which all developed economies have passed. The first is self-sufficiency, where the producer is his own market and there are as many outlets for agricultural produce as there are producers. The second state involves the existence and use of local markets and exchange facilities. By this time the number of outlets for agricultural produce has been considerably reduced. With increased specialisation of production, goods are sold to middlemen who act as distributors to the final consumers. With increased development of the market system farmers begin to outnumber dealers to a greater and greater extent. The third and final stage comes with the intervention of large food-processing industries and institutional buyers with great scale economies, which begin to dominate the agricultural markets. In 1964, 83 per cent of the total output of animals in the United States and nearly 40 per cent of the dairy produce were bought by food processers.[2] With one in four meals in the United States eaten away from home, the institutional buyer is also taking a more active part in agricultural purchasing. During this recent period of increase in size and decrease in number of outlets for agricultural produce, there has been little significant change in the number of farmers producing for market.

The growth of such large purchasing units has been reflected in the ever-increasing amount of preparation carried out by food processers and the decrease in home preparation of food. These large food firms are usually well informed and advised about the nature and state of their market, and because they are few and large they have considerable effects on current price levels of agricultural commodities.

Agricultural marketing problems can be summarised as:

1. There has been an ever-increasing concentration of ownership and control in the agricultural marketing and processing industries.
2. Production capability, through technological development, is outstripping the ability of the market to dispose of the extra production at an acceptable price.

3. Farmers are very price-sensitive. As they have a fixed area of production they are subject to fairly constant production levels. The only way in which incomes can alter is by price changes; both gross and net revenues are controlled directly by price.
4. With relatively small production units individual farmers have no control over price levels and over their own returns.
5. Demand is inelastic but yields may fluctuate considerably without any direct control by the producer.

For these reasons farmers traditionally believe that they are being wagged like the 'tail of the economic dog'. Bad production methods, or husbandry, are not usually the cause of fluctuating or declining agricultural revenues.

Two distinct solutions to these problems began to be evident in the period between the wars in the United States and in Britain. One solution involved the intervention of government by controlling market prices and improving market channels; the other solution involved the formation of a number of different types of cooperative organisation among farmers, sometimes with direct government assistance, in an attempt to emulate the marketing capabilities of their adversaries, the buyers. The nature of government intervention and the success of the cooperatives has varied greatly between different countries and indeed between different types of producer. The origin of both is to be found in the depressed state of agriculture in the period following World War I. In the United States agitation by farmers for a fair price policy started in the early 1920s. Net farm income in 1921 was only 39 per cent of its peak in 1919.[3] The prices paid to farmers fell substantially faster than the prices farmers had to pay for seeds, fertilisers and machinery. During the same period agriculture was similarly depressed in the United Kingdom because of what contemporary governments assessed as poor marketing circumstances. Channels of disposal were antiquated, circuitous and excessively costly and packaging and presentation of an essentially good set of products were also poor.[4]

The United States government reacted to this situation with the McNary-Hangen Bill of 1927. This bill introduced the concept of support prices for farmers to preserve income levels. On the domestic market, prices of agricultural commodities were to be fixed at defined 'fair' levels and all production which could not be sold at this price would be bought by the government and sold abroad at the best obtainable price. High tariffs would also be needed to protect the home producer from cheaper foreign imports. The losses, if any, incurred by the government in selling overseas would then be shared by all farmers in some system of collective taxation. Although this bill was twice vetoed because it was interpreted as an unwarranted interference in the operation of free enterprise, the principle of fair or support prices was firmly established and was to be the basis for government intervention in agricultural marketing in the United States for a considerable time.

The Agricultural Marketing Act of 1929 regenerated these policies with the establishment of the Federal Farm Board to undertake loan and storage operations. Farmers would be able to hand over those products which could not be sold at or above the support level for government storage and would receive a loan equal to the support price of the goods. If goods were eventually sold at or above the support price, the loan could be repaid. This policy was killed by the onset of the depression but was restarted by the Agricultural Marketing Act of 1933, the policies of which have remained in force to the present time for a number of commodities. A fair price was fixed according to a 1910–14 base level. This did not allow for such changing agricultural technology as the replacement of horses by tractors, so in 1948 the base price was fixed according to prices obtained over the preceding ten years. A Commodity Credit Corporation was established with two programmes:

1. For those products which farmers could store, for example grains, tobacco and cotton, a loan was available up to the support price. If the farmer could not sell at or above this price before the product started to deteriorate, the government bought it at the support price and repaid the loan. This material was then disposed of at the best available price in such a way that it did not depress the home market price. This normally involved selling abroad.
2. For non-storable products the Corporation was empowered to buy the goods at the relevant support price and then to attempt to sell them or alternatively to store them and finally, if necessary, to destroy them. This alternative system was applied to eggs, turkeys, butter and beef.

In order to return to a more normal market state, in 1948 support prices were fixed according to production levels. If it was a year of heavy production, the support prices fell accordingly and in years of shortage support prices rose. There is now mandatory support for corn, cotton, wheat, tobacco, rice, peanuts, dairy products, honey, tung nuts* and wool. Other products have had support prices applied from time to time depending on the state of the market and production levels.

This first type of government intervention was aimed at ensuring that farmers received a guaranteed return for their production. In order that this should not totally distort the economic pressures of supply and demand, some restrictions on levels of production were introduced, one type of which is an acreage allotment to each farmer restricting his area of planting for a particular crop each year. This is a voluntary control but the support price is payable only to those farmers who have agreed to the acreage control. In the case of severe over-production marketing quotas are introduced. The farmers affected can vote corporately to accept either the acreage controls, with heavy enforced penalties for breaking them, or a 'free for all' situation where farmers are free to grow all they wish, in which case the support price is lowered drastically to resemble more realistic market conditions. The advantage in these circumstances clearly lies with the farmer who can get his full production on to the market at the earliest possible moment, before the market price begins to fall in response to the over-production.

From the late 1950s onwards there was a shift in policies towards limiting production with the introduction of the National Soil Bank in 1958.[5] The soil bank had two objectives which were to some extent complementary: to reduce the production of surplus farm commodities which the government was finding it increasingly difficult to dispose of, despite considerable amounts given in foreign aid programmes, and to promote the conservation of the nation's land and soil resources. The latter goal is reached through the operation of the Conservation Reserve Programme which had its greatest impact in the areas of greatest soil erosion risk, particularly in the so-called dust bowl in the dry south of the Great Plains. The first objective has been sought by the establishment of an Acreage Reserve Programme through

* Used in the manufacture of a varnish drying agent, production of which is limited to the Gulf Coast states and to Florida.

which the farmers are paid for land on which crops are not grown: the amount depends on the normal use of the land. In 1958, for example, 18 dollars an acre was paid for wheat land taken out of production and up to 223 dollars for an acre of tobacco land. No crop may be harvested from these lands and no livestock grazed. This situation was aptly predicted by Joseph Heller in his novel *Catch-22* (1955, p 25):

> His speciality was growing alfalfa, and he made a good thing out of not growing any. The government paid him well for every bushel of alfalfa he did not grow. The more alfalfa he did not grow, the more money the government gave him, and he spent every penny he didn't earn on new land to increase the amount of alfalfa he did not produce. Major Major's father worked without rest at not growing alfalfa.

The effects of the Acreage Reserve Programme have been considerably more widespread than that of the Conservation Reserve.

This most recent policy of the United States government has shown a well-marked shift away from direct price interference in the market to a policy of restricting production and thereby controlling prices.[6] One of the reasons for this shift is to be found in a number of pessimistic government reports on the effects of acreage allotment programmes.[7] Although these policies had some effects on the level of carry-over of commodity stocks from one year to the next, the general conclusion of these reports was that controlling acreage does not effectively control production. The reduction of the acreage allowed in certain crops led to a concentration of fertilisers and effort on to the allotted acreage, and this was in turn concentrated on the most suitable land (Table 23). Partly because of better weather conditions, support

TABLE 23 Change in harvested area and yield for selected United States crops, 1953–5

	Harvested area per cent of 1952 area	*Yield per hectare per cent increase over 1952*
Wheat (1953–5)	−30	15
Cotton (1953–5)	−31	28
Corn (1953–5)	− 1	1
Rice (1954–5)	−28	16

(After United States Department of Agriculture, 'The effects of acreage allotment program 1954 and 1955: a summary report', *Production research report*, 3 (1956))

prices for cotton could have been between 8 and 15 per cent lower in 1955 than in 1953 in areas of cotton acreage control, without any loss of income to farmers. However, without acreage controls farmers would have grown between 20 and 56 per cent more cotton. The same pattern was repeated for wheat, while the acreage control of corn affected only that sold directly for cash, not for feed. One of the side effects of this policy was a considerable increase in the acreage of non-controlled crops, particularly feed grains as these were normally substituted for the controlled crops. Carry-over stocks* of feed grains increased considerably (Table 24). The solution involving the soil bank was to ensure that a proportion of the available acreage was taken totally out of production.

TABLE 24 United States carry-over stocks of selected crops, 1952–5

Crop	*Changes in carry-over stocks*			
	1952	*1953*	*1954*	*1955*
Wheat*	107·8	119·8	42·3	15·2
Corn*	99·4	53·2	38·4	42·6
Rice†	−2·27	22·22	101·61	31·75
Cotton‡	2·8	4·1	1·5	3·5
Feedgrains§	61·7	43·5	65·3	36·3

* $\times 10^7$ litres.
† $\times 10^6$ kilogrammes.
‡ $\times 10^6$ bales.
§ $\times 10^8$ kilogrammes.

(After United States Department of Agriculture, 'The effects of acreage allotment program 1954 and 1955: a summary report', *Production research report*, 3 (1956))

Another answer to the same problem of depression of agricultural prices was the establishment in the United Kingdom of marketing boards, especially of the Milk Marketing Board in 1933. The concept of a marketing board was laid down in the Agricultural Marketing Acts of 1931 and 1933. The assumption underlying these Acts was that existing low margins in agriculture were the result of deficiencies in marketing. It was argued that there must be scale economies to be gained by corporate marketing and that the newly established Milk Marketing Board with an annual turnover of more than a million pounds should be

* Carry-over stocks are those bought at support prices by government and which have not been disposed of by the end of the year.

able to afford the best possible advice and return the benefits derived to the producers. The intention was to remove the disparity of bargaining power of the small producers *vis-à-vis* the large and well-organised purchasers and distributors. Boards established for milk, eggs, potatoes, wool and some horticultural crops were revived after the war. During the war the tables had turned to some extent, for, with a great reduction of imported food, home suppliers found themselves in a sellers' market and a number of controls, especially rationing, were necessary to protect the consumer. After the war the function of the marketing boards changed. They became largely vehicles for the payment of subsidies and the operation of price-support policies much along the lines of the policies in operation in the United States at the same time. It has been suggested that farmers benefited far more by the action of the boards, or from the commodity commissions established in parallel with them, in their actions of price support than in any reorganisation of marketing.[8]

With the removal of food rationing in the 1950s no further boards were created although there was considerable pressure from farmers for the creation of a fat-stock marketing board.[9] For commodities which did not have a marketing board many of the same functions were provided by a commodity commission and by research units designed to improve agricultural production methods and marketing. For example, the Pig Industry Development Unit was established in the 1950s and there were a number of similar schemes for horticultural crops[10] and cereals. The Home Grown Cereals Authority was established under the Cereal Marketing Act of 1965 particularly to introduce a forward contract bonus scheme in an effort to reduce the importation of barley and to encourage the use of United Kingdom supplies;[11] although established to ensure an orderly market, its main function was the distribution of bonus payments.[12]

Although marketing boards have been established all over the world,[13] mostly because of the necessity of some form of control during two world wars, the concept of a marketing board is epitomised in the British Milk Marketing Board. Throughout Europe and north America the dairy industry is dominated by a very large number of small producers, with few purchasing and processing agencies, which are often government-run. Dairy produce is perishable and seasonal, so that the industry provides a microcosm of agricultural marketing as a whole. The major problem with the dairy industry is to ensure that, with relatively stable demand, production in winter, the low season, is sufficient to cover the demand for fresh milk. Surplus production in summer, a consequence of meeting the winter demand, has to be channelled into the manufacture of milk products which have a longer storage life than fresh milk. These products, made chiefly in the summer, are then marketed throughout the year. There is a further problem which the Milk Marketing Board has to tackle—the distribution of fresh milk and milk surpluses between different natural regions of the country where milk production and demand have relatively little geographical coincidence (Table 25).

TABLE 25 Regional distribution of milk supply and demand, England and Wales, 1966

*Region**	*% of milk production*	*% of population*
North	32·6	35·5
Midlands	20·4	20·0
South	12·7	34·4
South-West	22·6	5·5
Wales	11·7	4·6
Total	100·00	100·00

* For definition of these areas, see Milk Marketing Board, *The structure of dairy farming in England and Wales* (1965).
(Based on Straus, E., and Churcher, E. H. 'The regional analysis of the milk market', *Journal of Agricultural Economics*, 18 (1967), 221–40)

At the time of the foundation of the Milk Marketing Board there were nearly a quarter of a million farms in the United Kingdom of which over three-quarters were of less than one hundred acres in size and over one-half less than fifty acres. Production was very much in small-scale scattered units. The economic case for linking the marketing of these producers was overwhelming. The case for such a linkage being compulsory was also strong, on the principle that no one should receive benefits from a system without contributing to it.[14] The function of the Milk Marketing Board was intended to be mainly organisational, ensuring that all markets are fully satisfied all the time and involving the collection of milk in small quantities from producers, and its bulking and shipping, often over considerable distances, to areas of demand.[15] The establishment of a common marketing organisation, with checks and balances to ensure that no producers had special advantages over any others, has tended to spread the production of liquid milk throughout the country. Initially a pool price for milk was fixed for each of the eleven regions into which England and Wales are divided by the Board. This pool price was the result of the financial returns from all milk sold in the region: the higher the proportion of milk which went to a liquid milk market the higher would be the pool price. Thus regions with a high demand for liquid milk, notably the south-east, had a higher pool price than elsewhere. In order to stabilise prices over the whole country, and to ensure that boundaries between regions did not show undue price differentials, a system of regional compensation was introduced to ensure that the pool price per gallon never varied between regions by more than a very small amount. This variation now represents little more than 1 per cent of the total price. As a result, milk production in more distant areas, with greater reliance on the sale of milk for manufacturing, became increasingly attractive as the returns to the farmers were comparable to those available to farmers in the immediate vicinity of London. This breakdown of geographical locational advantage within the dairying industry has been further enhanced by charging farmers only for the transport cost involved in the movement of milk to the nearest depot of the Board. The shipment of such milk as is necessary over longer distances is paid for by the Board. These local transport rates contain differentials, the highest being charged to those producers at a greater distance from the main liquid milk markets and the lowest rates to those in proximity to these markets. Despite these differentials, the rates charged are still less than those actually incurred and the locational advantages to producers near markets are reduced (Table 26).

The pool price is derived from production of enough liquid milk to supply the national market. Because of the seasonality of production this is set at 25 per cent over the maximum production needed in the summer, with any further production sold for processing, thereby reducing the price paid to farmers. The structure of the pool price is indicated in Table 27. There are additional bonus prices to be achieved through production of high-quality milk with a total of twenty-eight payment price codes, three less than the basic price, one at the basic price, and the remainder over the basic price as a bonus for the production of milk with various butter-fat and non-fat solids contents. There

TABLE 26 Percentage changes in milk production by region, England and Wales, 1951–65

*Region**	*1951–6*	*1956–61*	*1961–5*
North	−1·3	−1·1	+1·6
Midlands	−0·4	−0·8	−0·4
South	−0·1	−0·5	−1·2
South-West	+1·1	+1·9	−0·2
Wales	+0·7	+0·5	+0·1

* For definition of these areas, see Milk Marketing Board, *The structure of dairy farming in England and Wales* (1965).
(Based on Straus, E., and Churcher, E. H. 'The regional analysis of the milk market', *Journal of Agricultural Economics*, 18 (1967), 221–40)

are also price reductions which are applicable on the basis of hygiene quality tests. In recent years increased milk production, with an increased proportion of milk sold for manufacturing (Table 28), has tended to lower the price at a time of increasing costs and this has reduced the number of producers by driving small-scale operators into other more profitable forms of agriculture, while the remainder increase the size of their herds. Government organisation of milk marketing has removed much of the dependence of milk production on distance from large urban markets for fresh milk, and recent increases in costs have made some producers move out of dairying. This change is most frequent in the south and east where there are alternative forms of

TABLE 27 The structure of the milk producers' pool price, England and Wales, April 1970–March 1971

	Pence perli tre
Government guarantee for the standard quantity (the amount considered adequate to ensure that the liquid market was fully supplied)	5·427
In excess of standard quantity guaranteed at the average market realisation of manufacturing milk.	2·465
Weighted average of these two prices (Effective guaranteed price)	4·867
Plus: MMB trading profits administration costs 0·034	
Minus: sales promotion costs miscellaneous charges −0·029 −0·055 −0·040	−0·09
Gross amount available for producers' prices	4·777
Capital levy payable by producers	−0·032
Net amount received by producers	4·745

(Source: Milk Marketing Board)

TABLE 28 Destination of total off-farm milk sales, England and Wales, April 1970–March 1971 (millions of litres)

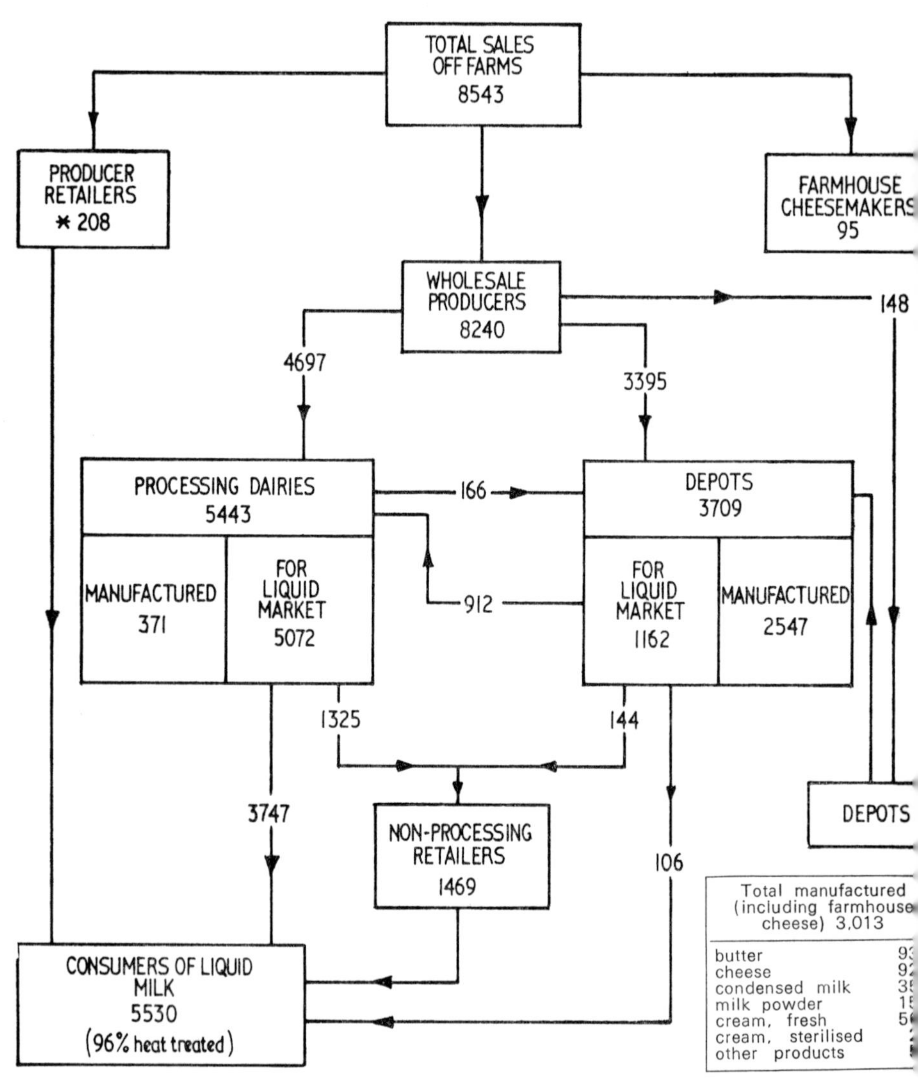

* Includes farm-bottled milk sold to other distributors and not direct to consumers but excludes milk purchased from the Board to supplement producer retailers' own production.

(Source: Milk Marketing Board)

NB. There are small rounding errors in this table resulting from conversion to metric units.

production, and has been less common in the west and north. The work of the Milk Marketing Board, therefore, tends to reverse the situation expected in a free market as envisaged by von Thünen. Milk production continues to increase in the more distant areas of the west of the country while it is replaced by alternative forms of agriculture in the areas adjacent to the main markets in the south-east.

The Milk Marketing Board is very much a producers' organisation with fifteen of its eighteen members elected on a regional basis by producers and the remaining three appointed by the Minister of Agriculture. The votes of producers are weighted so that each has one personal vote plus one more for each ten cows or part of ten cows in his herd. Although initially established by the government, the Board was always intended to act on behalf of producers, primarily in the marketing field. The Board has also adopted other functions, mainly of an advisory nature; for example, advising on breeding and herd management problems and on all questions of farm management. The Board also operates quality control programmes and advice. Outside its main function of milk marketing, however, the Board's greatest activity is in the area of milk manufacturing. The distribution of creameries (Figure 39) reflects the pattern of milk production surplus to local liquid milk requirements (the percentage production by region used for all types of milk manufacturing is shown in Table 29). Milk products manufactured by the Milk Marketing Board are distributed and sold by the Board under its own brand name. An increasing function of the Board has been in sales promotion and advertising, partly because of its own interest in manufactured goods and partly because of increased liquid milk production.

However, despite its title, the main effort of the Board, in common with other such boards, has been in the operation of fixed prices although its effects in the introduction of different marketing methods have been considerable, especially the introduction of bulk milk collections. The chairman's report for 1971 emphasises the situation with regard to prices and costs when he states that over the last fifteen years producer prices have risen by 7 per cent while factor prices have risen by up to 50 per cent. This is a repetition of the sort of complaint that gave rise to the first government intervention in agricultural marketing in the 1930s.

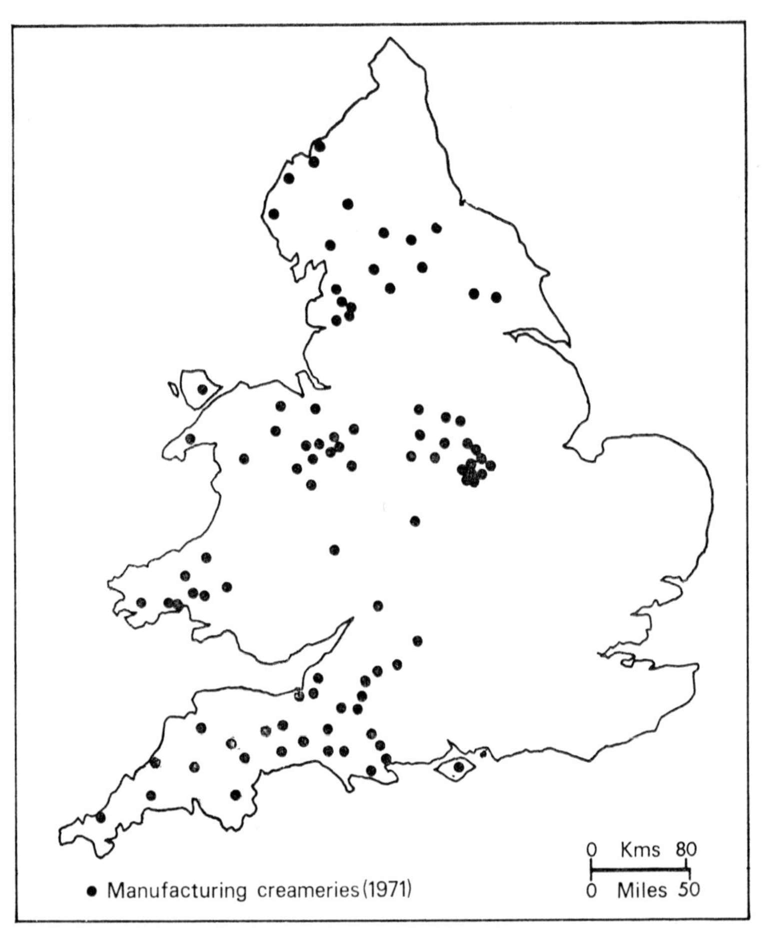

Fig 39. Location of manufacturing creameries in England and Wales, 1971 (excludes fresh creameries). (After data from the Milk Marketing Board planning department, 1971.)

TABLE 29 Milk production by region, England and Wales, 1969–70 (October–September)

Region	Milk production Total (1000s, litres)	Percentage of total production
1. Northern	402,236	14·7
2. North-Western	296,263	10·8
3. Eastern	14,932	0·5
4. East Midland	44,674	1·6
5. West Midland	373,844	13·7
6. North Wales	120,711	4·4
7. South Wales	333,197	12·2
8. Southern	45,783	1·7
9. Mid-Western	522,875	19·1
10. Far-Western	529,135	19·3
11. South-Eastern	52,804	1·9
Total for all regions	2,736,454	100

(Source: Milk Marketing Board)

DAIRY MARKETING IN THE UNITED STATES

The marketing circumstances of the dairy industry in the United States are not very different from those in Britain. Seasonal production, the need to satisfy a winter level of demand for fresh milk, the disposal of surplus summer production for processing and a recent rise in the degree of specialisation among dairy farmers with a reduction in the number of small spare-time farmers and an increase in the number and size of large herds—all are features of United States dairy production. The solution adopted was not to establish a milk marketing board but to apply Federal Milk Marketing Orders, each applying to a milk distribution area, with no single agreement for the whole country. With the great diversity of environments within the United States it would have been difficult, and perhaps undesirable, to establish a single marketing policy for the whole country. By 1964 approximately half the total volume of milk sold in the United States was sold under Federal Marketing Orders. These Orders were begun in the 1930s to stop distributors lowering prices excessively to gain distribution monopolies at the expense of the suppliers. A 'blend' price is fixed, depending on the percentage of the total production in the area sold as liquid milk or going to make various manufactured products and on the current prices of these alternatives. The price to all producers is based on the butter-fat content of the milk produced and if a producer's milk is used for manufacture at a relatively low price he is not penalised for this. Individual producers can have little control over the use of their milk and without these Marketing Orders they would have little control over their income. The blend price does something to stabilise a farmer's income but nothing to even out the seasonal nature of the supply. Local areas adopt one of three solutions to encourage winter production:

1. The blend price is varied according to the season depending on the percentage of the production which goes towards liquid milk sales. This has the effect of raising winter prices and encouraging production in the season of short supply.
2. The adoption of a base/excess plan. A base level of production is fixed for the winter months and excess over this base, produced during the summer, receives a lower price.
3. The fall premium plan. Deductions are made from summer payments to farmers and these are pooled in order to pay production incentives during the winter.[16]

In addition to these pricing regulations the Federal Government also has some control over the total production volume by imposing quality control based on the butter-fat levels of the milk. Producers who do not reach the desired levels cannot market their milk and if this happens consistently they naturally tend to change to other forms of farming. Despite these marketing controls, with a substantial proportion of total milk sales being outside the control of Federal Marketing Orders, price stabilisation and market interference is at a lower level in the United States than in Britain.

A more extreme case, where a marketing board is operating mainly as a method of administering price and production controls is the British Potato Marketing Board. Potatoes have a highly inelastic demand situation and, in times of high yield, prices cannot fall sufficiently to ensure that all available produce is sold. In these circumstances there has to be an alternative method of disposal. All growers with more than an acre planted to potatoes have to register with the Board. The Board does not buy all the production direct from the farmer as in the case of the Milk

Marketing Board, but restricts production by allocating acreages to each producer which he is forbidden to exceed. If these restrictions are not sufficient, and there is still a market glut in seasons of high yields, the Board may support the price by buying potatoes. If there remains an unsaleable surplus these may be bought by the Board and sold for stock feeding. In many cases the potatoes do not leave the farm, but are bought from and sold to the same farmer. There are over 50,000 registered producers and 3,500 registered merchants authorised by the Board to deal in potatoes. These merchants organise the distribution of the crop from the producers to the markets, either for direct home consumption or, for over 40 per cent of the production, for manufacture into potato crisps, dehydrated potatoes and other products.

INTERNATIONAL MARKETING POLICIES

Price support and marketing policies are not confined to individual governments. In 1958 at the time of the creation of the European Economic Community, one of the most pressing problems in Europe was the state of agriculture, especially its uneven economic development among the constituent countries of the Community. A common agricultural policy was adopted to increase productivity, to provide a fair standard of living for all farmers, to stabilise seasonal and annual market conditions and to ensure stable and reasonable retail prices. By early 1962 a common policy had been adopted for marketing grain, pigs, meat, poultry, eggs, fruit, vegetables and wine.[17] EEC prices were protected by the imposition of tariffs on imported goods raising their price to near the highest of the home prices. Price support operates in a rather different way from that adopted in Britain or the United States. The European Agricultural Guidance and Guarantee Fund operates the price support, not by making direct payments to farmers but by buying in the open market at any time that the price falls below established norms. These purchases are then stored or sold outside the community, more often than not at a loss. Money for support buying is gained through national levies, with Germany the largest contributor and France the largest receiver (Table 30). High prices have encouraged the production of greater surpluses which, without continuing support, would further reduce prices. In 1968 four thousand

TABLE 30 Agricultural levies in the EEC, to December 1968

Belgium	− 61	million dollars
France	+439	million dollars
Germany	−370	million dollars
Italy	−107	million dollars
Luxemburg	− 4	million dollars
Netherlands	+103	million dollars
Total	542	million dollars

+ = receipts from the Guarantee Fund
− = payments to the Guarantee Fund
(Source: Clout, H., *Agriculture: studies in contemporary Europe*, 1971)

million kilogrammes of wheat were surplus to sales and had to be stored by the Fund. This represents nearly 12 per cent of total wheat production.[18] The 'butter mountain', consisting of 406 million kilogrammes in 1969, although considerably reduced now, is another illustration of how price support, in conditions of stable demand, leads to over-production if the policy is operated without acreage or production restraints. In 1969 the bill for the purchase, storage and destruction of surplus production was over two million dollars. In a European context, with a very large number of small peasant farmers (in 1968 only 3 per cent of the farms were of more than fifty hectares and three million farms had less than five cows), a more fundamental solution to the problem is required. This needs to involve widespread social reform including an accelerated drift from the land and considerable increases in the farm size. In 1971 the Community adopted a joint policy of social reform and made available 1,500 million dollars for this purpose. The emphasis of this policy is to reduce the farming population by providing pensions for older farmers and setting up retraining facilities for the young. Prices but not farming returns will be lowered by increased efficiency and this will stimulate an increased demand. This solution is sought not through production controls but by stimulating the market through reduced prices.

Britton,[19] discussing the problems of surplus grain production particularly in the United Kingdom, suggests ten methods by which this surplus could be reduced. These can be divided into two categories: reduced production, and increased sales. The policies described could be applied to any other area of surplus production, and they therefore form a useful summary of the scope of possible government interference in agricultural production and marketing.

Reduced production

1. Make break crops in cereal rotation schemes more profitable by new or extended subsidies; there is, for example, already a subsidy on field beans, often used as a break crop.
2. Substitute wheat and maize for barley production; of these two maize is more promising and has very recently begun to achieve considerable acreages in south-east England.
3. Reduce the guaranteed price or support price; this would certainly reduce production, but other forms of income support for farmers would probably be required.[20]
4. Impose a full system of production quotas.

Increased sales

5. Encourage exports, either by direct subsidy or by aids for shipment, storage, etc.
6. Increase domestic consumption, particularly in the area of stock feeding. The fluctuating demand for feeding materials, depending on the weather, particularly in early spring, is one of the main reasons for fluctuating demand for cereals in the United Kingdom.
7. Reduce imports to make disposal of the home surplus easier. United States grain is imported direct to Northern Ireland and French grain was on sale in the United Kingdom in 1968 at about half the domestic price in France.
8. Improve the quality of grain produced, especially by reducing its moisture content, thereby improving home demand.
9. Find, and direct surplus to, new uses for home-grown grain.
10. Establish and control a system of carry-over stocks to iron out fluctuations in production. At present no subsidy is paid for grain which has not been sold off the farm by 21 July of the year following the harvest, thus making the establishment of carry-over stocks unlikely.

In summary, then, there are three approaches adopted by governments to control the free play of agricultural markets. The most important and the most widespread is price fixing with usually some form of government purchase of surplus production. As an improvement and addition to this policy there is acreage control thereby limiting production. Finally there are government efforts to improve marketing channels and thereby to raise returns to the farmers. In some cases all three functions are performed by marketing boards, although as we have seen in the case of the British Milk Marketing Board and the Potato Marketing Board, the first two approaches usually dominate.

COOPERATIVE ORGANISATIONS

It is difficult to define a cooperative in the case of agricultural production and marketing. In the strict sense a cooperative should involve the total sharing of all expenses and returns, whereby the weak are supported and the strong are kept at an average level of return. A more typical level of agricultural cooperation is found when a group of farmers combine to work as a larger unit for production, or for aspects of production, purchasing, processing or marketing. Such cooperatives need not involve equal treatment of all partners but may be run on more strictly commercial lines. Distinctions within the range of cooperatives are difficult to make as they can occupy any position on the whole range from a true cooperative to a group of friends helping each other with the harvest.

Whatever the degree of cooperation, the groups are designed to enable the farmers to reach backwards towards their suppliers and to so reduce the inequality of numbers and bargaining power between themselves and their suppliers, or to reach forward, with the same objectives, towards their markets. As Hewlett[21] has pointed out, interest in cooperative movements is inversely proportional to the level of contemporary agricultural prosperity. Thus, in the United States cooperatives first appeared in the early 1900s but they reached their peak in the depression following World War I. In Sweden a national structure of cooperatives was established in the 1930s organised on a local, regional and national level. Practically every farmer in Sweden is a member of at least one such cooperative organisation, with the dairy industry wholly cooperative and fatstock production and marketing up to 75 per cent cooperative in organisation and eggs up to 65 per cent.

Cooperatives are generally most successful in areas of bulk production of single crops. In the United States they have been most successful in the north-central states for the marketing of grain and also in areas of citrus fruit production. In 1960 it was estimated that 60 per cent of citrus production was marketed through cooperative organisations, 58 per cent of the butter and 50 per cent of the grain crops.[22]

In the United Kingdom, although there are a number of schemes for group buying and production, marketing cooperatives have grown relatively slowly and now account for only some 5 per cent of the total produce handled. The lack of success is probably due to the actions of the government price support and the promotion of marketing boards coupled with the relative prosperity of British farming since the last war. A further factor must be the heterogeneous nature of British farming. There are few areas of bulk production of single crops such as are common in the United States. Production is mixed and the farming population is socially and economically diverse, two factors which make the implementation of full cooperation unlikely. On the other hand, the British government has recently openly encouraged the formation of production cooperatives as an alternative to restructuring agricultural operations. To reduce production costs by restructuring the resource inputs to the individual farm is difficult because of the immobility of land and capital.[23] This reallocation is made easier if a number of farmers can be persuaded to operate their farms as a single unit. The 1967 Agriculture Act established the Council for Agricultural and Horticultural Cooperation with resources for grant aid. This body has had most success in areas of large arable farms in eastern England but despite this direct government encouragement such cooperatives are few in number. In contrast cooperative bulk buying and selling, as represented by such organisations as the Eastern Counties Farmers, is very extensive. Whatever the state and degree of marketing cooperation involved, the purpose is clear; it is a private-enterprise solution to the problems of agricultural marketing and low returns thought to be a consequence of the low bargaining power of individual farmers. In terms of their success in raising farm incomes it is doubtful whether they are as effective as government price support and acreage control policies.

CONTRACT GROWING

A third type of reaction to the agricultural marketing situation is the introduction of contract production and marketing. This solution is normally initiated by the food processing industries rather than by producers and it involves an attempt by those industries to achieve a measure of vertical integration by moving backwards towards the production end of their business and exerting a measure of control. This can be necessary for a number of reasons concerning quality control, regularity of supplies, and a full integration of the factory process with the harvesting of the crop. For certain crops, especially vegetables for quick freezing, this integration is of the utmost importance. Very often contract production is closely linked with cooperatives or production groups, where the contract arrangements are negotiated by a separate company established by the farmers for this purpose. Again strength is gained in the bargaining position by corporate representation.

Perhaps the most extreme example of contract production is provided by the United States broiler chicken and egg industries.[24] In this case the purchasing company may provide the chicks, feed, vaccines, fuel, litter, advice and general supervision. The farmer is left with little more than the role of caretaker of an industry which happens to be using a portion of his land.[25] The same sort of position can be found in pig breeding. The feed company may provide the weaners and market the fat pigs on the producer's behalf. In these circumstances any loss is borne by the producer while he has no control over the health and quality of the weaners with which he is provided. Such a contract is hard for the farmers and is not by any means typical. Contract farming can be useful to the farmer, especially if he is just starting in business or is changing to a new line of production. Risk is shared with the contracting company, capital requirements of the farmer are reduced considerably and, provided there is enough profit in the arrangement, it may allow the farmer to accumulate capital to achieve eventual independence. On the whole, there are few economic criticisms which can be raised against contracting.[26]

There are, therefore, certain circumstances in which a contract between a farmer and a purchaser is desirable or even essential. These are generally cases where control over production is desirable, either to limit total production or to link production closely to factory processing capacity. Sugar beet marketing in the United Kingdom provides an example of contract growing to limit production to the capacity of the sugar beet processing plants. It is obviously inefficient to grow more beet than can be processed, and it is even more wasteful to grow so little that some of the available plant is under-utilised. It was partly to control this situation that the British Sugar Corporation was established by the Sugar Industry (Reorganisation) Act of 1936. The first sugar beet processing plant to be more than briefly successful was at Cantley in Norfolk, established in 1921, by which date factories were numerous throughout Europe. Britain did not have sufficient incentive to establish sugar processing facilities based on beet because of her sugar-producing colonies.[27] By 1936 there were fifteen companies operating in five groups, each of which differed considerably in its production policy. The location, and still more the size, of the plant differed widely depending on the history and origin of the controlling group. For example, the Anglo-Dutch groups of companies, which included the plant at Cantley, were noticeably larger than others because the Dutch already had experience with large-capacity plants. The problem was to select the right size of plant to gain from economies of scale and yet not be so large as to be unable to obtain sufficient beets from the local area. Since the Reorganisation Act the same seventeen plants have remained in operation, with the addition of one at Cupar in Scotland, each supplied more or less by its immediate area (Figure 40). Most production is confined to a radius of 50 kilometres, although loads in excess of 10,000 kilogrammes can be transported by rail and the grower is reimbursed for the costs incurred over and above the Corporation's rate for 64 kilometres.

Beet is grown under contract, with each grower restricted to a limited acreage. This restriction is necessary as sugar beet grows well in eastern England and the crop is useful not only as a rotation root crop but also because the tops can be used for feed and growers can be supplied with factory waste products usable as further cattle feed. In addition there is a growers' subsidy. The contracts are necessary to ensure planned production geared to factory capacity. The result of these policies is to even out the production of sugar beet over the whole of eastern England, and the locational advantages of immediate proximity to the pro-

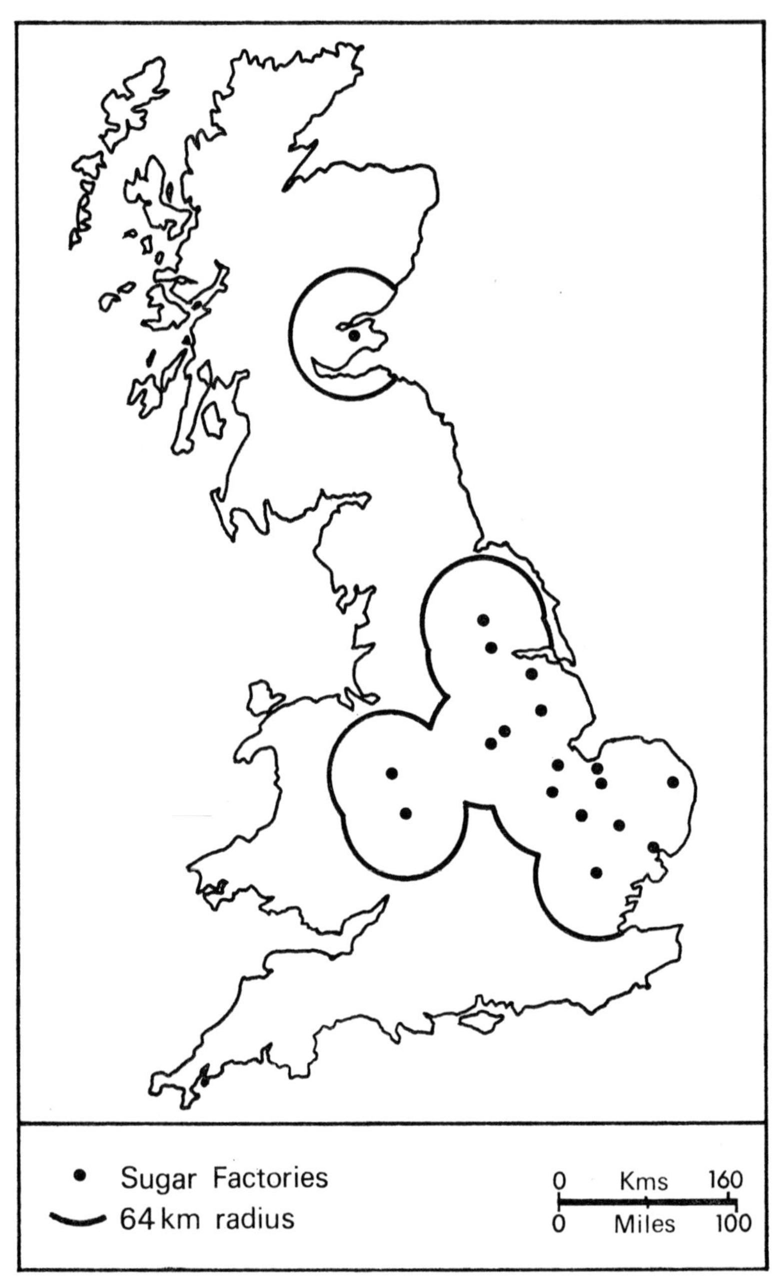

Fig 40. The location of sugar beet factories in the United Kingdom and their main catchment areas. (After Coppock, J. T., *An agricultural geography of Great Britain*, 1971.)

cessing plant has little effect on the distribution of growing areas. For example, although there are three sugar beet processing plants in Norfolk, the distribution of sugar beet (Figure 41) shows little tendency for the crop to be concentrated in proximity to these plants. Rather production is concentrated on the loams of north-east Norfolk and on the silts of the southern Fenland while the crop is little grown on the heavy clay lands in the south of the county.

Most vegetable freezing plants in the United Kingdom are located on the eastern seaboard, based originally on eastern ports when fish are landed. With a rapidly expanding market for frozen foods of all types, these plants have diversified to become multi-product freezing operations. With a highly seasonal production of vegetables, these plants need a diverse base and many freeze all sorts of green vegetables, potatoes, and meat products as well as fruit. On the other hand, because of their port location and the requirement to freeze many products within a short time of harvest, their location is far from ideal. Only half of the potential catchment area is available because of the sea; moreover the roads to the coast are often highly congested in summer, contracting still further the potential catchment area. Because of rapid expansion in the market for frozen food, recently established factories have been located inland as, for example, at Spalding and Grantham.[28]

Producers started growing vegetables for freezing on an individual basis, adapting their existing machinery to deal with field vegetables. Steadily the advantages of cooperation among farmers became evident to both the producers and to the manufacturers. Two types of cooperative became established: one deals with the production side while the other, often involving some of the same people, is concerned with harvesting. Two contracts are involved, one for each of these functions. It is necessary to separate harvesting contracts, involving considerable outlays of capital for specialised equipment, to ensure that the harvesting pattern can be directly under the control of the manufacturer, independent of producers, and can be geared exactly to his plant freezing capacity. Contracts are arranged by the company with grower groups and with individual farmers who are not members of such groups. In areas of large, heavily capitalised farms, grower co-operatives are not common,[29] but approximately 80 per cent of the production for factories in Yarmouth and Lowestoft comes

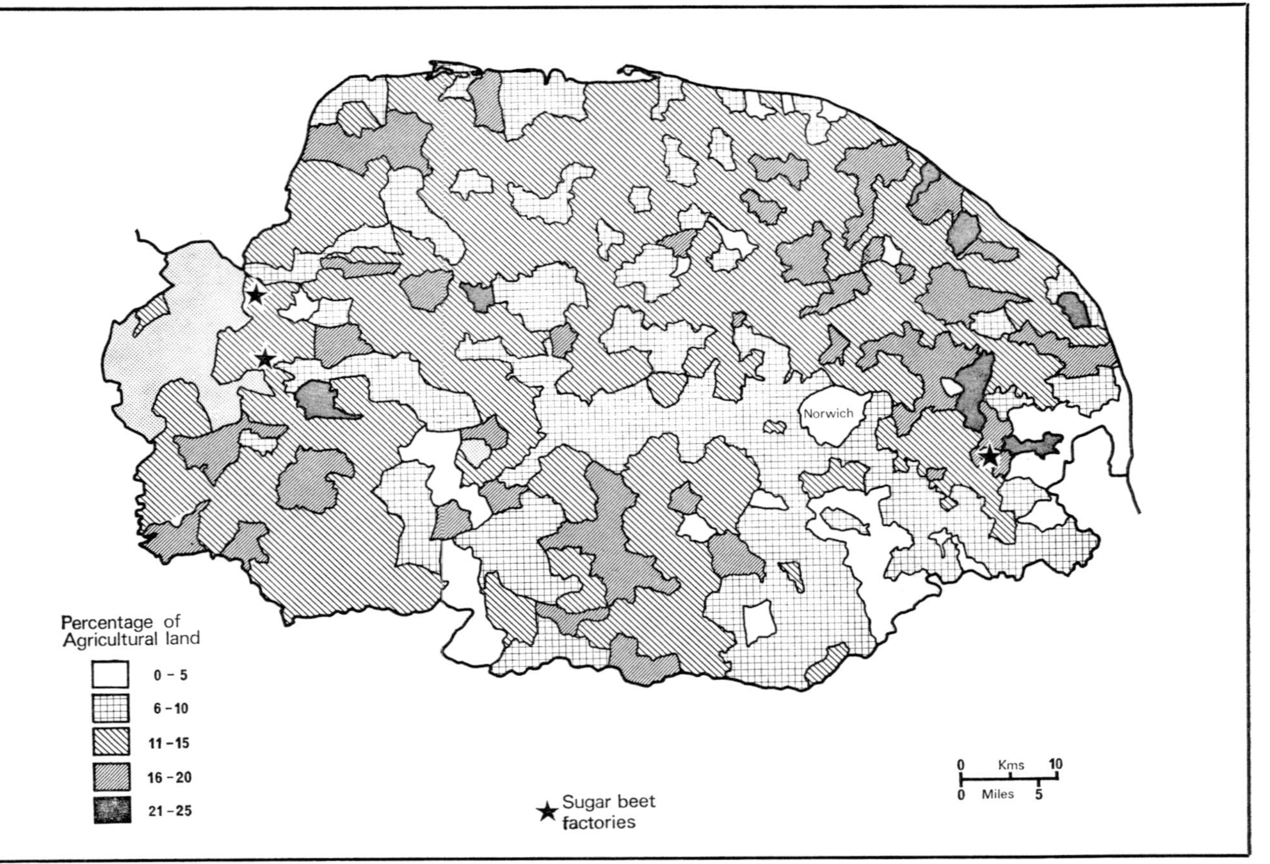

Fig 41. The distribution of sugar beet by parish in the county of Norfolk. (Source: Ministry of Agriculture, Fisheries and Food, agricultural data, 1968.)

from such groups in East Anglia. The company designates the land on which the crop will be grown, the acreage to be allotted to each farmer, and, most importantly, the date of planting. The company also recommends sprays and fertilisers but actually controls only the date of application of insecticide to ensure that a sufficient period before harvest follows the last application. Payments to farmers depend on quality and yield but may be shared according to acreage in some groups, or varied according to yield in others. The choice of the distribution of returns is left to the grower cooperatives and forms a part of the contract.

Harvesting is the area of greatest control, and this in turn controls the area over which vegetables can be grown for freezing. A rule is operated which stipulates that all peas to be used for freezing have to be quick-frozen within ninety minutes of harvest. No land more than sixty-five minutes' drive by road from the factory can grow peas. This allows fifteen minutes to harvest enough peas to fill a vehicle, and ten minutes' quality control and weighing at the factory. Smaller vehicles are used for transport from the greater distances to shorten the loading time in the fields. This production area could be expanded by ensuring a greater harvesting capacity to enable the lorries to be loaded more quickly, or by installing small field freezing plants divorced from the main factory. Neither of these solutions is economic for the large companies, although remote freezing operations are conducted by some smaller companies to extend their catchment areas and to increase their turnover of single products.

CONCLUSION

No attempt has been made in this chapter to discuss the working of individual agricultural markets in any detail. Other sources can be used to discover the nature of the market for different agricultural commodities.[30] Rather we have isolated a number of factors in agricultural marketing which give rise to special problems and have seen how these are tackled in a number of different ways in western economies. The problems which stem from the existence of a large number of producers, in competition with a limited number of purchasers, are met by governments by the encouragement of marketing boards and privately by various degrees of cooperation among producers. Surplus production, leading to market glut conditions, and lowering of returns to agriculture, is met by various government price support policies and restrictions on production. In addition, certain special circumstances may encourage the development of contract production, although the present overall importance of this for agricultural marketing is limited.

CHAPTER EIGHT

Competition for Agricultural Land

AGRICULTURAL land use in Europe expanded, though not continuously, from the period of the beginnings of settled agriculture up to the end of the nineteenth century, by which time the contemporary economic margins of production had been reached and indeed in some cases surpassed. The total agricultural land in Britain has since been reduced from two directions. First, the twentieth century has seen some retreat from these margins, while, secondly, there has developed an increasing level of competition with other uses of land. Most of the competing uses for land are directly or indirectly associated with urban growth. This reduction in agricultural area is regarded as more or less dramatic depending on the viewpoint of the investigator. However, in terms of agricultural production, the loss of agricultural land has been more than mitigated by a substantial improvement in output per hectare during the same period. The possibilities of increasing productivity to compensate for further loss of agricultural land, as well as catering for a growing population of increasing affluence, is a matter of some debate, not only in the United Kingdom but on a worldwide basis. In the United Kingdom, up to the year 2000 there is little doubt that such a trade-off is possible, although this must depend on the acceptable overall quality of the environment.[1]

There are three elements in the calculation of a land budget, which can be defined as availability of land on the one hand balanced by the requirements of agricultural production from that land on the other. The first and most publicised element is the loss of agricultural land to other uses, especially but not exclusively to urban uses. Of considerable importance in the consideration of this element is not simply the area of land lost but also the quality of the land. The second element concerns the change in population numbers and the economic standards and demands of that population. The third element involves changes in the production techniques of agriculture in order to increase yields on the lessening agricultural area and the possibilities of developing new lands for agricultural use. These topics have been studied elsewhere[2] and they involve a consideration of population dynamics, changing dietary habits as well as plant breeding and development. This chapter concentrates on only one aspect of the problem, that of a diminishing resource of agricultural land available in developed countries to supply food for a growing population.

THE CONVERSION OF LAND FROM AGRICULTURAL USE

Land is lost to agriculture in a variety of ways in an expanding industrial economy, perhaps most obviously to urban uses. This change of land use takes place in highly populated areas, witnessed by large numbers of people and affecting them directly by providing land for new houses. However, conversion to urban use is far more extensive than that for housing alone as it involves the related phenomena of schools, roads, docks, storage and warehouse facilities, industrial uses, shopping centres, water supply reservoirs and sewage disposal works and many more. In addition, as we shall see, the spread of urban growth into a rural area involves a substantial area not yet in urban use but in immediate juxtaposition with the town, which is substantially blighted or even sterilised for agriculture by impending change and by other kinds of urban intrusion. Moreover this conversion process is virtually irreversible; land once converted to urban use is most unlikely ever to return to any form of agricultural production.

Urban expansion is largely, but not wholly, the result of population growth. Predicting future changes in population is of the utmost importance for assessing the future life of a 'resource' like agricultural land just as it is for any other resource. However, urban demands for land would continue to rise in present circumstances even with stable populations. The fact that the urban fabric of cities in Europe and America is outgrowing its original functions and yet continues to survive, in many cases unsatisfactorily, means that this fabric has to be continually replaced. It will be replaced at the current standards rather than the standards in operation at the time of building. The present United Kingdom space requirements for housing, including gardens, demand that agricultural land be used to rehouse large sections of the existing population without taking account of any population increase, either from natural change or by migration. As an example, the number of persons per room in England and Wales has fallen considerably since the beginning of the century (Table 31), and, with the present trend to smaller families in somewhat larger houses, this increase in the demand for housing space can be expected to continue. Similarly, households' demand for water has continued to increase with better opportunities for personal hygiene in the form of flush toilets and fixed baths (Table 31). With increased affluence other domestic water demands also rise. Taking into account the rapidly rising industrial demands for water, there is a continued loss of agricultural land for reservoirs. Valley reservoirs rarely use poor-quality agricultural land and so the acreage loss to agriculture is really an understatement of the effective loss.

Recreation is rather less directly a consequence of urban expansion, but none the less a considerable rival for agriculture in the battle for the use of land. Some types of recreation may not be incompatible with agriculture, but often agriculture is severely affected, if not completely removed by the competition. Golf courses can use very poor quality agricultural land but are normally near built-up areas where such land may not be available. The use of land for caravan sites, motor racing tracks and other similar activities are an obvious sign of land used for recreational purposes but more pervasive is that of casual recreation, based usually on the family car, which uses the countryside in an unorganised and substantially unmeasurable way, though its effects on good-quality farm land and farming must be considerable. Damage to crops, worrying of animals by family dogs and the problems of litter are but three examples of intrusion

TABLE 31 Population densities, England and Wales

Percentage of population of England and Wales living at different densities

	Persons per room				
	>1½	1–1½	1	½–1	<½
1911*	25·0	23·2	15·0	22·0	9·7
1961	7·0	14·5	19·3	45·8	13·3

Percentage of households of England and Wales living at different densities†

	Persons per room				
	>1½	1–1½	1	½–1	<½
1961	3·8	8·5	16·4	47·0	24·2
1966‡	1·6	4·9	59·5		33·9

Private households in England and Wales without exclusive use of fixed bath and water closet

	Bath	*WC*
1951	37%	8%
1966‡	15%	1·4%

* This row does not add to 100 per cent. The remainder of the population do not live in private households.

† Comparison of household and population density shows how large families live at high densities.

‡ Data based on 10 per cent sample, census, 1966.

without total loss of agricultural land. Similar types of intrusion occur at the peripheries of built-up areas where urban land is juxtaposed with rural land use. As the urban area expands so does the length of this contact zone between two basically incompatible land uses. The loss to agriculture by the conversion from rural to urban land uses is made considerably larger if this intrusion factor is included.

THE AMOUNT OF LAND CONVERSION

Several studies have been made of loss of agricultural land to urban and other uses.[3] Since 1945 there has been an average land loss of 15,600 hectares per year in England and Wales but, although it might be thought that losses to agricultural land would have increased, actual losses were highest between the two wars with about 24,500 hectares lost between 1927 and 1928 and between 1938 and 1939. Anticipated future losses to urban land will be between 18,200 and 19,200 hectares per year for Great Britain, or about 1 per cent per decade of the agricultural land.[4] As this loss will be concentrated in the south and east of the country it is likely to be of better-than-average quality and may well involve about 7·5–8·0 per cent of the country's total agricultural potential. Losses to forestry, at an anticipated 1·2 per cent of the available agricultural land per decade, will have a much smaller effect on agricultural production as forestry will occupy some of the poorest land. If forestry turns to the production of more hardwoods, which are not suited to the wet mountainous areas of the west of the country where most of the present forestry activity is concentrated, further loss of agricultural potential can be expected.

Estimates for land loss to urban use in the United States vary considerably but the figure is certainly over 400,000 hectares per year[5] and is accelerating fast.[6] With a tradition of low density living where housing regulations in the outer suburbs may require a density of one house for every three or more acres the loss of agricultural land will be very considerable and higher per annum in proportion to the population than is experienced in Europe. On the other hand, because of the size of the country, it is only around the major metropolitan areas, particularly on the east coast, that this agricultural land loss is considered seriously as a threat.

THE PROCESS AND EFFECTS OF LAND CONVERSION

The process and effects of conversion from rural to urban uses can best be seen in terms of the factors controlling the demand for rural land on the one hand and the factors controlling the supply of land suitable for conversion on the other. Where the demand is at least partly met by the supply there are a number of characteristic effects. We have already considered demands on rural land for urban use to be the result of growing urbanisation of the world's population. In the United States in 1850 only 6 per cent of the population were living in cities of over 100,000 population. This percentage had increased to 15·5 by 1890 and to 62 per cent by 1960 and continues to grow rapidly.[7] Comparable figures for other countries are shown in Table 32. Although the proportion of the total population involved varies considerably from one part of the world to another, the phenomenon of rapid urbanisation is virtually universal.

TABLE 32 Growth of cities, 1800–1960 (percentage of total population living in cities of 100,000 or over)

Country	*1800*		*1850*		*1890*		*1960*
England and Wales	9·73	(1801)	22·58	(1851)	31·82	(1891)	60·2
Scotland	0	(1801)	16·9	(1851)	29·8	(1891)	55·8
USSR	1·4	(1820)	1·6	(1856)	3·2	(1885)	23·9
United States	0		6·0		15·5		61·8
France	2·8	(1801)	4·6	(1851)	12·0	(1891)	26·4
Prussia	1·8	(1816)	3·1	(1849)	—		—
Germany (East and West)	—		—		12·1		43·9
Italy	4·4		6·0	(1848)	6·9	(1881)	25·8
Spain	1·45	(1820)	4·4	(1857)	6·8	(1887)	29·2
Switzerland	0	(1822)	0		0	(1892)	31·5
Belgium	0	(1800–10)	6·8	(1846)	17·4		42·5
Netherlands	11·5	(1795)	7·3	(1849)	16·6	(1889)	42·5
Denmark	10·9	(1801)	9·6	(1840)	17·3		38·9
Norway	0	(1801)	0	(1845)	7·6		21·8
Sweden	0	(1805)	0	(1850)	7·34		24·7
Portugal	9·5	(1801)	7·2	(1857)	8·8		20·7
Ireland	3·1	(1821)	3·9	(1851)	8·2	(1891)	30·4

(Source: Hoyt, H., 'The growth of cities from 1800 to 1960 and forecasts to the year 2000', *Land Economics*, 39 (1963))

A large urban population increase does not necessarily in itself lead to a rapid suburban development, since this population can be housed in a number of alternative ways some of which make little inroads on the agricultural land. A combination of circumstances is needed for rapid suburban development; Clawson[8] lists six factors considered important in suburban development in the United States since 1945:

1. Between 1945 and 1958 ten million new households were formed, partly as a result of a backlog of demand formed during the depression and the war and partly as a result of lowered marriage ages and greater marriage rate.
2. The lack of land suitable for family house building in existing urban areas and the high price of any available land.
3. The great flexibility which came as a consequence of the motor car and the affluence it represents. Although high prices, the necessity for car ownership and high costs of journey to work kept the middle and lower income groups out of the

expanding suburbs, this did not noticeably impede suburban growth because of rapidly rising levels of personal income.

4. Industrial work centres also became freed by motor transport to locate in the suburbs, especially on or near the new interstate freeway networks.
5. With movement of population and employment to the suburbs it was to be expected that retailing would follow. This development was accentuated by changing methods of retailing requiring large single-storey buildings and extensive parking lots.
6. The demand for new housing is not inelastic but is extremely credit-sensitive. Very few people have sufficient accumulated capital to avoid the necessity for credit when buying a house. A policy started during the New Deal to make mortgages more freely available through Federal guaranteed insurance programmes did much to encourage new house-building. As a result of these Federal schemes standards of house-building became almost universally applied throughout the country and systems of tax concessions on loans for house owners had also helped to encourage suburban development. Break[9] has shown that the liberalisation of credit terms in the United States in the period after 1934 had a small but significant effect on housing demand, an effect which became more marked after 1945. There is also some indication that these credit facilities enabled more money to be spent per house than had previously been the case. The relationships among some of these variables affecting housing demand are shown in Figure 42.

One important factor in explaining rapid suburban expansion in parts of the United States and Britain is the preference of suburban families for horizontal rather than vertical living. Postwar development in California, for example, was of a horizontal nature, partly because of the fear of earthquakes, but probably far more importantly because of the history of the Californian settlers, most of whom had spent at least some time living in the spacious middle west of the United States.[10] In addition the suburb settlements of California had no legacy of life without the motor car and so had no high-density models to copy for later developments.

No matter how great the forces of urban expansion, such expansion can take place only at the expense of the surrounding

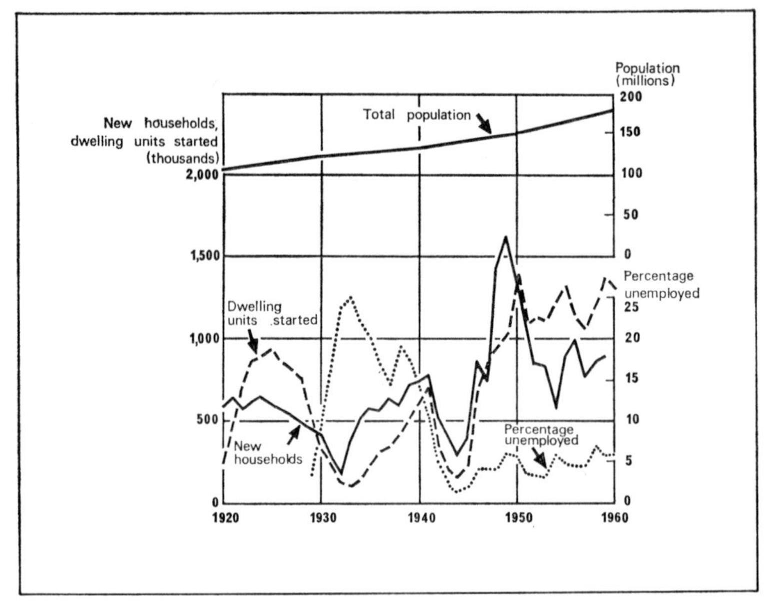

Fig 42. Relationships between population, unemployment, new households and dwelling units started in the United States between 1920 and 1960. (After Clawson, M., *Suburban land conversion in the United States: an economic and governmental process*, 1971.)

land. Expansion presupposes that this land is available in sufficient quantity for intending purchasers. Von Thünen's agricultural land-use model shows us that this land into which the settlements will expand will be used for some form of intensive production and we might therefore expect to see some level of resistance to its use for urban spread. The competitive situation between agricultural land use and urban land use is described by Muth[11] in terms of an extended von Thünen model. On the one hand there is a rapidly increasing demand for urban land, and on the other is an industry producing goods in a very inelastic demand situation. As agricultural improvement takes place more and more crops can be produced, but these cannot always be sold at a sufficiently high price to warrant their production. In such a situation the demand for urban land use must be more flexible than that for agricultural use and urban land will win the competition. Because of the different rent gradients of urban and rural land use, Muth calculates that increased demand for housing will have to be at least 1·4 times that for food to move the urban margin outwards. In situations of local supply of agricultural products like fresh milk, demand will be elastic: as the city grows so will the demand for food. The bid rent for agricultural uses will rise as well as that for urban uses and the margin will not move substantially (Figure 43). In cases where the surrounding agricultural land produces only some of the food requirements for the country as a whole the increase in the population of the city will have only a marginal effect on the demand for the locally grown food product and the city will expand at the expense of the rural land. Because of the different elasticities of the two products of the contrasting land uses, urban land will almost always expand at the expense of rural uses if the population of the city is expanding.

However, the effect on farmers is more subtle and complex than in this simple economic analysis. One of the most direct incentives for farmers living on the urban fringe to sell land for urban development is a negative change in their cash flow brought about by rising property taxes with little or no corresponding inflow of extra money. This situation does not apply in the United Kingdom as farmers do not pay property taxes on their land or their buildings provided they are used wholly for agricultural purposes. In the United States, on the other hand, as the urban area grows so more money is needed by local government to provide schools, roads and services which are needed primarily for the benefit of the new urban dwellers, but which are paid for from the local property taxation. Although the farmer's land will have increased in value owing to its proximity to the expanding urban centre, this is of little benefit unless and until the investment is realised. Blase and Staub[12] point out the basic inequality of treatment between a stockholder taxed on his interest, or income, and on his capital gain should he sell his holdings, and the farmer who is taxed both on his income and on the basis of his holding. In a number of sampled farms near Kansas City the authors found that this local tax loss was sufficient to encourage some farmers to sell their land. The detrimental effects of the property tax situation on farmers near the urban fringe was also noted in California by Griffin and Chatham[13] and Gregor.[14]

The rapidly escalating land prices on the city margin provide

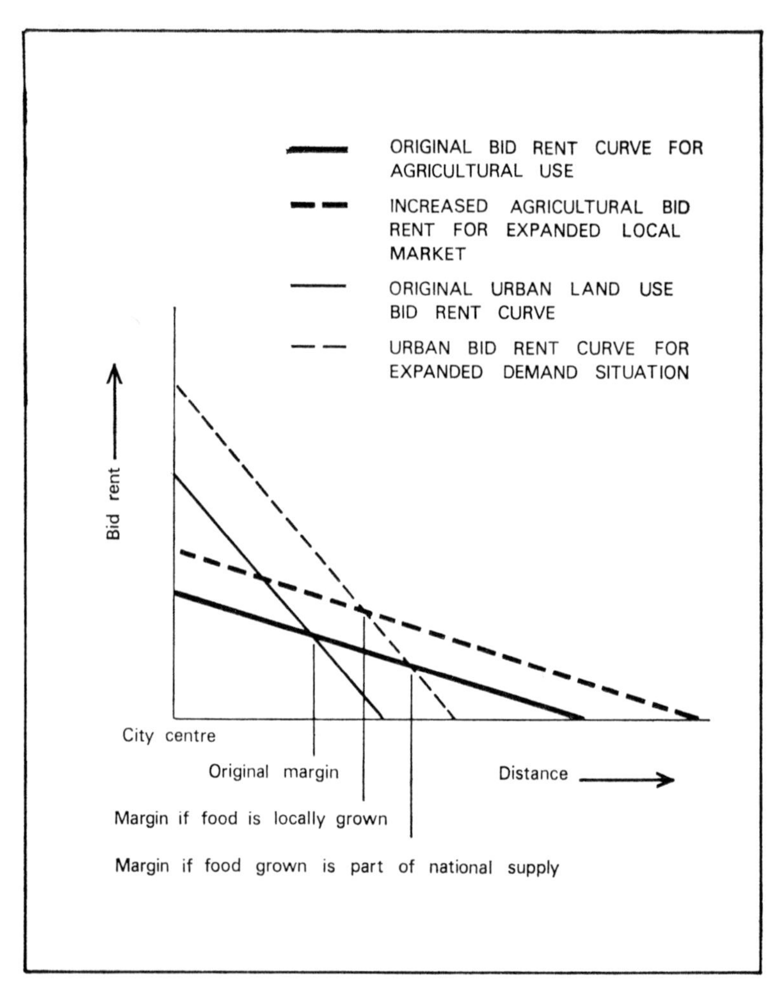

Fig 43. Bid rents for urban and rural land uses on city margins. If demand for rural and urban land increases the city margin will move outwards. If the demand for urban land increases and rural demand remains static the city margin will be pushed further from the city centre.

their own incentive for the demise of farming. Not only is there a temptation to sell and to reinvest in land further away from the city margin, but this temptation will become particularly important in cases where the agricultural practice requires periodic investments of capital and labour, as in, for example, fruit orchards where the trees will have to be periodically replanted.[15] Farmers may consider such reinvestment as of marginal benefit with high land values and high taxation, quite apart from any other form of interference brought about by the proximity of urban dwellers. Other proximity effects accelerate the decline of agriculture: crop spraying is impeded, pilferage increased, especially of soft fruits, and pollution from the city may decrease yields. Gregor[16] estimates that between 1953 and 1957 smog damage to agriculture in the Los Angeles area may have exceeded three million dollars. In addition, the water table may be substantially lowered by increased urban demands on water and by the built-over area and street drainage systems limiting infiltration and ground water recharge. In areas where agriculture requires ground water for irrigation this lowering of the water table may be another factor requiring a substantial investment of capital to rebore or to deepen the wells.

THE TRANSFER OF LAND

For a number of reasons it is not usually the farmer who sells his land directly for urban development. Well in advance of the urban 'front' is what has been called a shadow zone where land is purchased and hoarded for later use. There are a large number of potential sellers each of whom has to offer only a relatively small parcel of land. The purchasers of this land, who may be regarded as speculating on the chances of its eventually increasing in value, are acting as a type of land bank accumulating plots large enough for eventual sale for building.[17] As it is held by speculators, such land is unlikely to be used for agriculture except where it is of very exceptionally good agricultural value, when it may be rented and worked out with little or no investment in fertiliser and improvement. Although each piece of land left temporarily idle may be relatively small it is of considerable value and in aggregate can amount to a very substantial proportion of the total land in and around the city. In 102 of the cities of over 100,000 population in the United States in 1966, 22·3 per cent of the land was in private hands and idly awaiting future development. Northam[18] identifies five major types of idle land in and around cities:

1. Small, irregular plots left by developers as unsuitable for building. This will be especially important where the local zoning or planning requirements stipulate a minimum size of housing plot and where the area of land available for development is no simple multiple of such a minimum size.
2. Large, but physically unsuitable plots, perhaps liable to flooding or on steep slopes.
3. Land held as corporate reserve, built up by local companies as a hedge against future land scarcity and inflation of land prices.
4. Speculation land, awaiting open sale for development when the price is right. Planning permission may have been granted and services may be available.
5. The institutional reserve of land held in public ownership for future development of schools and other services.

The total amounts of such land in United States cities of different sizes are shown in Table 33: of this land about 78 per cent could be considered as suitable for building in the future.

TABLE 33 Idle urban land in United States cities, 1970

Idle land of all types in cities :	
over 1 million population	8·7% of total area
between 500,000 and 1 million	23·9% of total area
between 250,000 and 500,000	18·6% of total area
between 100,000 and 250,000	27·4% of total area
86 out of the 130 cities over 100,000 population	19·7% of total area

(Source: Northam, R. M., 'Vacant urban land in the American city', *Land Economics,* 47 (1971), 345–55)

A further type of idle land is described by Wehrwein[19] in an early discussion of the problems of the rural-urban fringe. Where urban expansion has produced an over-rapid outward growth of residential sub-division, with pavements, roads and often other services running through otherwise vacant land, the land is subdivided and the plots sold and often resold many times through speculation; as a result ownership becomes confused. This type of urban speculation has been typical of the urban frontier in the United States, starting in the river towns to the west of the

Appalachians and continuing to California and Florida.[20] Land is thus forced prematurely out of agricultural use and the land in von Thünen's innermost ring becomes as idle as that in the outermost, too distant from the city for agriculture to repay the cost of transport to market. Such development is now largely controlled by zoning orders in the United States and by planning controls in the United Kingdom. Land left completely idle is rarer in the United Kingdom although the productivity of agriculture is reduced in the rural-urban fringe.

THE RESULTS OF RURAL-URBAN LAND-USE CHANGES

Surrounding cities is what Boyce[21] has called a precession wave with considerable turbulence in land use, land values and land ownership. It is a zone characterised by an intermingling of land use, involving an irregular transition from farm to non-farm land. There is also much speculation in land and much agricultural land falling into disuse. Such a wave moves ahead of the main urban advance and may be considered to be absorbed by it. The direct effects on agriculture result from one of two extremes. On the one hand land may to all intents and purposes leave agriculture for all time, while on the other hand on some of the remaining land the proximity of a large and expanding urban market encourages such intensive forms of agriculture as market gardening and dairy production. Such agricultural areas are characterised by a smaller number of farm workers than in totally rural areas and by a larger number of part-time farmers.[22]

The extent to which this loss of agricultural land can be said to have serious effects on food supplies both locally and nationally depends on the local circumstances. In California, where only about one-fifth of the total land area is suitable for agriculture and where that land is of high quality and subjected to a very large population pressure, the effects can be considerable.[23] The towns in the Santa Clara valley south of San Francisco developed as service centres for the agricultural community which had settled on the best agricultural land. As a result, with a rapid population increase, these towns have had to expand at the expense of the best-quality land. It can be argued that any shortage of suitable agricultural land will be accounted for in the operation of the land market. As land becomes scarce so agricultural prices will rise and more land which is currently idle will be taken into agricultural production because a greater investment of capital is made worthwhile by the higher rates of return now possible for agriculture. Agricultural production will then be able to compete with urban growth in the open land market. Gillies and Mittelbach[24] argue that land should be used for that function for which it has the highest value. There is, however, considerable doubt about the extent to which the free market can cope with this type of situation. There are many examples of almost total resource depletion where the market has operated to the long-term detriment of mankind.

The existence of agricultural shortages arising from the necessity of feeding more people on less land depends on the race between population and technology. The extent to which agricultural yields can be improved and new sources of food tapped to keep pace with population growth and increased living standards is debatable.[25] Almost all projections have taken a rather parochial view. Clawson[26] argues that there is plenty of land within existing United States suburbs for at least another generation of suburban dwellers and he estimates a demand for between sixty and sixty-five million new household units between 1960 and 2000. Thirty-nine million of these will be new units, the remainder making up replacement of ageing dwellings within the present area of the city. Similarly Edwards and Wibberley[27] are optimistic about the United Kingdom's ability to produce a major proportion of its own food on shrinking land resources over a similar period of time. However, with an increasing interdependence between nations it is shortsighted to consider either the United Kingdom or the United States in isolation. On a worldwide scale the problem agriculture has to face in providing for a rapidly increasing population would appear to be a great deal more acute and immediate, although an examination of this topic would go far beyond a consideration of the process of change from rural to urban land uses.

Conclusion

THE preceding eight chapters have been very wide-ranging and yet they have by no means embraced the full scope of agricultural geography. In particular there has been no more than passing reference to the description of different agricultural types and regions. Such accounts can be found elsewhere and many have been referred to throughout the text. It has often been said that geography lacks suitable theories and methods of analysis which can form alternatives to regional study. Many such methods of analysis have become available in recent years for geographical studies and their use has provided a suitable environment for further theoretical progress. The lack of theory providing a framework for explanation of spatial variation is particularly acute in agricultural geography where remarkably little progress has been made in the century and a half since the writings of von Thünen.

There are three main approaches to the theoretical study of agriculture. The first, through a straightforward model of geographical determinism, assumes that the physical environment controls agricultural decision-making. The second approach is similar and could be called economic determinism. This type of model assumes that the economic factors of the market, production and transport costs operate on uniform producers who react in a uniform and rational manner to these economic circumstances. The third type of model recognises that there is a further set of influences on agricultural decision-making which are not based on economic or physical environmental factors. Farmers' sets of values, identifiable aims and attitudes to risk are all important in final decisions concerning agriculture. In all these methods of approach, except that of physical determinism, the decisions taken by the farmers feed back to the decision-making environment, altering it and providing continuous change. Each of these approaches has been developed in isolation from others whereas the decisions taken by the very large number of producers in agriculture are the result of the operation and interaction of all three sets of circumstances. The relative importance of each will vary from individual to individual. Future progress in agricultural geography will be made in integrating these three contrasting methods of approach and incorporating feedbacks within the system to produce a more comprehensive, dynamic model of agriculture. This book, by concentrating on four main areas of interest, can provide a level of understanding about the present 'state of the art' to supply a platform for further theoretical work.

The first area of interest is a discussion of the existing models of agricultural location, derived from the early work on rent theory by Ricardo and von Thünen. The shortcomings of such rent theories are all too obvious but they form a considerable start to the formidable problems of providing explanatory models of agricultural activity. Whereas many refinements and modifications have been made to these models such changes have produced no radical departures from the deterministic partial equilibrium approach. The growth of interest in behavioural studies in geography has come about partly as a result of dissatisfaction with such economic determinism.

Instead of new models, or perhaps forming a necessary prerequisite for these, much progress in agricultural geography has been made in devising methods for the analysis of agricultural data, particularly in relation to the difficult problem of change with time as well as with space. Dynamic elements of agricultural geography can be studied by linear programming and Markov chain techniques and simulated with Monte Carlo methods. It is possible that such work will soon form the basis of a more dynamic treatment of agricultural location so that future models will be able to incorporate spatial and temporal changes in agricultural distributions and farm structure.

There has been a recent tendency in geography to have relatively powerful techniques of analysis but for the data to which such techniques are applied to be inadequate. All methods by which data are analysed have specific requirements; even the calculation of simple averages requires that the data are either interval or ratio. As the techniques increase in sophistication so do the demands on the nature of the data. Being aware of the problem in statistical analysis of 'rubbish in—rubbish out', an important section of this book has been devoted to the availability and reliability of agricultural data. As with technique development, data availability is constantly changing both in space and in time. Remote sensing techniques applied to agricultural data-gathering are undergoing particularly rapid development. With the advent of the Earth Resources Technology Satellites and Skylab and with the development of the space shuttle planned to be in operation in the 1980s, the potential for gathering large volumes of data about the occupation of the earth's surface has dramatically increased. This potential to gather data has tended to outstrip the overt demand for it and there have been several recent important examples of the technology of space-borne remote sensing chasing potential users of such technology rather than the more usual, and indeed more satisfactory, situation of potential users demanding an improved technology. However, groups of interested people are getting together throughout the world to identify uses of such a flood of new data which will take advantage of its synoptic character and repetitive nature. Monitoring of crop ripening on a continental, or even world, scale serves as an example of an application which seems quite feasible in the near future and the impact which such monitoring would have on world commodity markets cannot be underestimated. Unfortunately, not only has the data-collection technology overtaken the demands of potential users but it has also seriously overtaken the technology needed to use these data efficiently. Again progress is being made in devising methods for automatic scanning of this great potential data mass but we are far from sure that we can fully utilise the data that present remote sensing systems produce. No matter what the measure of success of remote sensing systems they will be able to collect only data pertaining to land use and there will remain a need for parallel improvements in the availability of data pertaining to farm management and profitability. Such data are collected at present by sampling and published only in a crudely composite form of limited value to the study of agricultural geography.

There are two particular aspects of agriculture which have been included only in rudimentary form in models of agricultural location: the role of the market and competing, non-agricultural, demands for agricultural land. The market is normally considered only as a 'sink' of limited capacity which is able to absorb agricultural produce at a given price. We have seen in the section of this book devoted to agricultural marketing that this role is far more complex. In particular the fact that there are many farmers producing for market and demanding that their governments do something to stabilise prices and thereby their source of income means that the present role of agricultural markets in many countries is far from the simple method of expression of supply and demand which the early deterministic models required. Although much has been learned of the complexities of agricultural markets, and the role of governments in this sector, little of this has been incorporated in agricultural location models and this provides one important area for future work. The pressures exerted on farmers by expanding urban populations and concomitant demands for land on which the farmer grows his crops cannot be fully developed within the framework of a simple rent model. Firstly the realm of the farmer and that of the urban dweller are not entirely separate and although there is usually a clear-cut boundary between the two land-use types the impact of the urban dweller on the rural environment is considerable well before the land use changes. In addition, although the processes of land conversion are well studied, non-economic and planning controls on land conversion which limit the operation of supply and demand are lacking from agricultural land-use models.

The more traditional role of agricultural geography has not been ignored. The importance of the region in geography, and in agricultural geography in particular, is well known and different techniques of regional division have been outlined. It is also important that geographers appreciate that there are other ways of ordering and simplifying data. There are continuous models which may describe data more accurately and usefully in terms of a steady change between extremes rather than allocating all to a series of distinct classes. Such models may eventually help with the widely recognised problem of the identification and justification of regional boundaries. As such data analysis methods, whether classificatory or continuous, are ways of simplifying and structuring data these discussions follow the outline of various data sources in agricultural geography. Having accumulated the data and subjected them to some form of preliminary analysis, further analysis is required to investigate the inter-relationships which make up agriculture. Such inter-relationships can be studied by simple visual methods using distribution map overlays but there are also a number of powerful statistical techniques which can be useful. Remembering the necessity to have data of a standard worthy of these techniques, something of what they can achieve has been outlined, as have some of the commoner difficulties in their application.

Taken as a whole, this book has attempted to steer a middle road between existing works on agricultural systems and regions and studies of pure theory of agricultural location. To obtain a full understanding of agriculture it is essential to have a wide appreciation of the available sources of data and the ways of analysing data as well as the ways in which practising farmers contribute to agricultural production. With the prospects of an expanding gap between the prosperous nations and the poor being expressed in a growing gap between food production and food requirements, an understanding of the agricultural sector of the economy, encompassing both production and distribution, will have potential benefit. If a predictive model of agricultural activity can be produced, agricultural geography, both descriptive and theoretical, will have reached maturity.

NOTES

CHAPTER ONE

1 Bunge, W. 'Theoretical geography'. *Lund studies in geography, Series C, General and mathematical geography*, No 1 (1966)

2 For example, Gourou, P., *The tropical world* (3rd ed, 1971)

CHAPTER TWO

1 Ricardo, D., *Principles of political economy and taxation* (1817)

2 Clark, C., 'The value of agricultural land', *Journal of Agricultural Economics*, 20 (1969), 1–23

3 Von Thünen, J. H., *Der isoliente Staat in Beziebung auf handwirtschaft und Nationalökonomie* (Rostock, 1826). Translated as Hall, P. (ed), *von Thünen's Isolated State* (Pergamon, 1966)

4 Chisholm, M., *Rural settlement and land use* (Hutchinson, 1962)

5 Clark, C., op cit

6 Chisholm, M., op cit

7 Alonso, W., 'A theory of the urban land market', *Papers and Proceedings of the Regional Science Association*, 6 (1960), 149–57

8 Hoover, E. M., *The location of economic activity* (McGraw-Hill, 1963)

9 Johnson, H. B., 'A note on Thünen's circles', *Annals of the Association of American Geographers*, 52 (1962), 213–20
10 Chisholm, M., op cit
11 Peet, J. R., 'The spatial expansion of commercial agriculture in the nineteenth century: a von Thünen interpretation', *Economic Geography*, 45 (1969), 283–301
12 Hoover, E. M., op cit
13 Grotwold, A., 'Thünen in retrospect', *Economic Geography*, 35 (1959), 346–55
14 Johnson, H. B., op cit
15 Chisholm, M., op cit
16 Ibid
17 Jonasson, O., 'Agricultural regions of Europe', *Economic Geography* (1925), 277–314
18 Van Valkenburg, S., and Held, C. C., *Europe* (2nd ed, Wiley, 1952)
19 Peet, J. R., op cit
20 Isard, W., 'A general location principle of an optimum space economy', *Econometrica*, 20 (1952), 406–30
Garrison, W., and Marble, D. F., 'The spatial structure of agricultural activities', *Annals of the Association of American Geographers*, 47 (1957), 137–43
Dunn, E. S., *The location of agricultural production* (2nd ed, Univ. of Florida, 1967)
21 Garrison and Marble, op cit, 137
22 Isard, W., op cit
23 Harvey, D. W., 'Theoretical concepts and the analysis of agricultural land use patterns in geography', *Annals of the Association of American Geographers*, 56 (1966), 361–74
24 Day, R. H., and Tinney, E. H., 'A dynamic von Thünen model', *Geographical Analysis*, 1 (1969), 137–51
25 Williams, K., *Linear programming: the simplex algorithm* (Longmans, 1969)
Baumol, W. L., *Economic theory and operations research* (2nd ed, Prentice-Hall, 1965)
26 Heady, E. O., and Egbert, A. C., 'Regional programming of efficient agricultural production patterns', *Econometrica*, 32 (1964), 374–86
27 Howes, R., 'A test of a linear programming model for agriculture', *Papers of the Regional Science Association*, 19 (1967), 123–40

28 Lentnek, B., Patten, G. P., and Jones, R. C., 'A spatial production function analysis of corn yields', *Proceedings of the Association of American Geographers*, 2 (1970), 85–8

29 Rae, A. N., 'Profit maximization and imperfect competition: an application of quadratic programming to horticulture', *Journal of Agricultural Economics*, 21 (1970), 133–40

30 Banzini, L., 'The difference in the south', *Encounter*, 105 (1962), 7–17

31 Von Neumann, J., and Morgenstein, O., *Theories of games and economic behaviour* (Princeton Univ. Press, 1944)

32 Langham, M. R., 'Game theory applied to a policy problem of rice farmers', *Journal of Farm Economics*, 45 (1963), 151–63

Agrawal, R. C., and Heady, E. O., 'Application of game theory models in agriculture', *Journal of Agricultural Economics* (1968), 207–18

Pillon, J. L., and Heady, E. O., 'Theories of choice in relation to farmer decisions', Agricultural and home economics experimental station, *Iowa State University Research Bulletin*, 485 (1960)

Gould, P. R., 'Man against his environment: a game-theoretic framework', *Annals of Association of American Geographers*, 53 (1963), 291–7

Hazell, P. B. R., 'Game theory—an extension of its application to farm planning under uncertainty', *Journal of Agricultural Economics*, 21 (1970), 239–52

33 Davenport, W., 'Jamaican fishing: a game theory analysis', *Yale University Publications in Anthropology*, 59 (1960), 3–11

34 Gould, P. R., op cit

35 McInerney, J. P., 'Maximin programming—an approach to farm planning under uncertainty', *Journal of Agricultural Economics*, 18 (1967), 279–90

McInerney, J. P., 'Linear programming and game theory models—some extensions', *Journal of Agricultural Economics*, 20 (1969), 269–78

36 Agrawal, R. C., and Heady, E. O., op cit, 207

37 Langham, M. R., op cit

38 Agrawal, R. C., and Heady, E. O., op cit

39 Downs, R. M., 'Geographic space perception: past approaches and future prospects', *Progress in Geography*, 2 (1970), 65–108

40 Coppock, J. T., *An agricultural Geography of Great Britain* (Bell, 1971)

41 Freeman, T. W., *Ireland, a general and regional geography* (4th ed, Methuen, 1969)

42 Coppock, J. T., op cit

43 Brown, L. A., 'Diffusion processes and location: a conceptual framework and bibliography', *Bibliography series No 4, Regional Science Research Institute* (1968)

44 Hägerstrand, T., *Innovationsförloppet ur Korologisk Synpunkt* (Gleerup, 1953). Translated as Pred, A., and Haag, G., *Innovation diffusion as a spatial process* (Chicago Univ. Press, 1967)

45 Marble, D. R., and Nystuen, J. D., 'Measurement of community mean information fields', *Papers and Proceedings of the Regional Science Association*, 11 (1963), 99–110

46 Bowden, L. W., 'The diffusion of the decision to irrigate', *Univ. of Chicago, Research Papers* No 97 (1965)

47 Rogers, E. M., *The diffusion of innovations* (Free Press of Glencoe, 1962)

48 Anderson, G. B., 'Innovation i landbruget', *Tolumandsbladet*, Copenhagen, 41 (1969), 186–9

49 Griliches, Z., 'Hybrid corn and the economics of innovation', *Science*, 132 (1960), 275–80

50 Ryan, B., and Gross, N. C., 'The diffusion of hybrid seed corn in two Iowa communities', *Rural Sociology*, 8 (1943), 15–24

51 Coleman, J., Katz, E., and Menzel, H., 'The diffusion of an innovation among physicians', *Sociometry*, 20 (1957)
Morrill, R. L., 'The shape of diffusion in space and time', *Economic Geography*, 46 (1970), 259–68

52 Pred, A., 'Behaviour and location', *Lund Studies in Geography*, Series B, 27 (1967)

53 Charlton, P. J., and Thompson, S. C., 'Simulation of agricultural systems', *Journal of Agricultural Economics*, 21 (1970), 373–90

54 Donaldson, G. F., and Webster, J. P. G., *An operation procedure for simulation of farm planning: a Monte Carlo method* (Wye College, Dept of Agricultural Economics, 1968)

55 Brown, L. A., 'On the use of markov chains in movement research', *Economic Geography*, 46 (1970), 393–403

56 Stevenson, G., *Mathematical methods for science students* (Longmans, 1961)
57 Drewett, R., 'A stochastic model of the land conversion process: an interim report', *Regional Studies*, 3 (1969), 269–80
58 Krumbein, W. C., 'Fortran IV computer program for simulation of transgression and regression with continuous-time markov models', *Univ. of Kansas, State Geological Survey, Computer Contribution*, 26 (1968)
59 Drewett, R., op cit
60 Colman, D. R., 'The application of markov chain analysis to structural changes in the north-west dairy industry', *Journal of Agricultural Economics*, 18 (1967), 351–61
61 Power, A. P., and Harris, S. A., 'An application of markov chains to farm type structural data in England and Wales', *Journal of Agricultural Economics*, 22 (1971), 163–77

CHAPTER THREE

1 Horscroft, P. G., 'Changes envisaged in the agricultural census for England and Wales', *Statistical News*, 6 (1969), 9–10
2 Best, R. H., and Coppock, J. T., *The changing use of land in Britain* (Faber, 1962)
3 Horscroft, P. G., op cit
4 Coppock, J. T., *An agricultural atlas for England and Wales* (Faber, 1964)
5 Horscroft, P. G., op cit
6 Jones, W. D., 'Ratio and isopleth maps in the regional investigation of agricultural land occupance', *Annals of the Association of American Geographers*, 20 (1930), 177–95
7 Coppock, J. T., 'The parish as an agricultural unit', *Tijdschrift voor Economische en Social Geografie*, 51 (1960), 317–26
8 Coppock, J. T., 'The relationship between farm and parish boundaries', *Geographical Studies*, 1 (1955), 12–26
Coppock, J. T., 'The parish as an agricultural unit', *Tijdschrift voor Economische en Social Geografie*, 51 (1960), 317–26
9 Best, R. H., and Coppock, J. T., op cit
10 Horscroft, P. G., op cit, paragraph 6.9

11 Board, C., 'Field work in geography, with particular emphasis on the role of the land-use survey', Chapter 10 in Chorley, R. J., and Haggett, P., *Frontiers in geographical teaching* (Methuen, 1967)

12 Stamp, L. D., and Willatts, E. C., *The land utilization survey of Britain: an outline description of the first twelve one-inch maps* (London School of Economics, 2nd ed, 1935)
Stamp, L. D., *The Land of Britain: its use and misuse* (Longmans, 1948)

13 Willatts, E. C., 'Changes in land utilization in the south-west of the London basin, 1842–1932', *Geographical Journal*, 82 (1933), 515–28

14 Scott Report: *Report of the committee on land utilization in rural areas* (HMSO, Cmd 6378, 1944)

15 Coleman, A. M., 'A new land-use survey of Britain', *Geographical Magazine*, 33 (1960), 347–54
Coleman, A. M., 'The second land-use survey; progress and prospect', *Geographical Journal*, 127 (1961), 168–86

16 Stamp, L. D. (ed), 'Report of the commission on inventory of world land-use', *International geographical union, 18th International Congress* (1956)

17 Jones, W. D., and Sauer, C. O., 'Outline for field work in geography', *Bulletin of the American Geographical Society*, 47 (1915), 520–5
Jones, W. D., and Finch, V. C., 'Detailed field mapping in the study of the economic geography of an agricultural area', *Annals of the Association of American Geographers*, 15 (1925), 148–57

18 Sauer, C. O., 'Mapping the utilization of the land', *Geographical Review*, 8 (1919), 47–54

19 Sauer, C. O., 'The education of a geographer', *Annals of the Association of American Geographers*, 46 (1956), 287–99

20 Trefethen, J. M., 'A method of geographic sampling', *American Journal of Science*, 32 (1936), 454–64

21 Proudfoot, M. J., 'Sampling with traverse lines', *Journal of the American Statistical Association*, 37 (1942), 265–70
Wood, W. F., 'Use of stratified random samples in land-use survey', *Annals of the Association of American Geographers*, 45 (1955), 350–67
Anderson, J. R., 'Towards more effective methods of obtain-

ing land-use data in geographic research', *Professional Geographer*, 13 (1961), 15–18
Haggett, P., and Board, C., 'Rotational and parallel traverses in the rapid integration of geographic areas', *Annals of the Association of American Geographers*, 54 (1964), 406–10

22 Berry, B. J. L., 'Sampling, coding and storing flood plain data', *Agriculture Handbook No. 237* (US Dept of Agriculture, 1962)
Berry, B. J. L., and Baker, A. M., 'Geographic sampling', in Berry, B. J. L., and Marble, D. F., *Spatial analysis: a reader in statistical geography* (Prentice-Hall, 1968), 91–100

23 Haggett, P., *Locational analysis in human geography* (Arnold, 1965), 197

24 Berry and Baker, op cit

25 Reid, I. D., *The application of statistical sampling in geographic studies* (Unpublished MA thesis, Univ. of Liverpool, 1963)

26 Board, C., op cit

27 Bennett-Jones, R., *The pattern of farming in the east midlands* (Univ. of Nottingham, School of Agriculture, Dept of Agricultural Economics, 1954)
Jackson, B. G., Barnard, C. S., and Sturrock, F. G., 'The pattern of farming in eastern England', *Occasional Papers No. 8* (Farm Economics Branch, School of Agriculture, Cambridge University, 1963)

28 Birch, J. W., 'Observations on the delimitation of farming type regions: with special reference to the Isle of Man' *Transactions of the Institute of British Geographers*, 20 (1954), 141–58

29 Others have been:
Burton, I., 'Types of agricultural occupance of flood plains', *Research Paper No. 75* (Univ. of Chicago, Dept of Geography, 1962)
Scott, P., 'Sample farms in northern Tasmania', *The Journal of Geography*, 2 (1963), 56–65

30 *National farm survey of England and Wales (1941–1943); a summary report* (HMSO, 1946)

31 Birch, J. W., op cit

32 Butler, J. B., 'Representative farms—a guide to decision making', *Journal of Farm Economics*, 45 (1963), 1449–55

33 Board, C., 'Land-use surveys—principles and practice', Chap-

ter 3 in *Land use and resources: studies in applied geography* (Institute of British Geographers, Special Publication No 1 (1968), 29)

34 Sisam, J. W. B., 'The use of aerial survey in forestry and agriculture', *Imperial Agricultural Bureau, Joint Publication No. 9* (1947)

35 Pestrong, R., 'The evaluation of multispectral imagery for a tidal marsh environment', *United States office of naval research, contract no. NONR-4430* (undated)

36 Board, C., 'The use of air photographs in land-use studies in South Africa and adjacent territories', *Photogrammetrica*, 20 (1965), 163–70
Brenchley, G. H., 'Aerial photography and agriculture', *Outlook on Agriculture*, 5 (1968), 258–65
Steiner, D., 'A world-wide survey of the application of aerial photographs to interpreting and mapping rural land use', in 'Aerial surveys and integrated studies', *Proceedings of the Toulouse conference* (UNESCO, 1968), 513–16

37 Pestrong, R., op cit

38 Colwell, R. N., 'Determining the prevalence of certain cereal crop diseases by means of aerial photography', *Hilgardia*, 36 (1956), 223–86

39 Cooke, R. U., and Harris, D. R., 'Remote sensing of the terrestrial environment—principles and progress', *Transactions of the Institute of British Geographers*, 50 (1970), 1–23

40 Fritz, N. L., 'Optimum methods for using infra-red sensitive colorfilm', *Photogrammetric Engineering*, 33 (1967), 1128–38

41 Pestrong, R., op cit

42 Cooke and Harris, op cit

43 Hoffer, R. M., and Johannsen, C. J., 'Ecological potentials in spectral signature analysis', in Johnson, P. L., *Remote Sensing in Ecology* (Univ. of Georgia Press, 1969)

44 Laboratory for agricultural remote sensing, 3 (1968), *Annual report* (Agricultural experimentation station, Purdue University, Lafayette, USA)

45 Sabins, F. F., 'Thermal infra-red imagery and its application to structural mapping in southern California', *Geological Society of America Bulletin*, 30 (1969), 397–404

46 Doverspike, G. E., 'Microdensitometer applied to land-use classification', *Photogrammetric Engineering*, 31 (1965), 294–306

47 Sabins, F. F., op cit
48 Cooke and Harris, op cit
49 Board, C., 'Land-use surveys—principles and practice', Chapter 3 in *Land use and resources: studies in applied geography* (Institute of British Geographers, Special Publication No 1 (1968), 29)
50 Carter, L. J., 'Earth resources satellite: finally off the ground?', *Science*, 163 (1969), 796–8
51 *Ecological surveys from space* (US National Aeronautics and Space Administration, 1970)

CHAPTER FOUR

1 Hartshorne, R., *The nature of geography: a critical survey of current thought* (Association of American geographers, 1939)
2 Whittlesey, D., 'The regional concept and the regional method', Chapter 2 in James, P. E., and Jones, C. F., *American geography: Inventory and prospect* (Syracuse Univ. Press, 1954)
3 Bunge, W., 'Theoretical geography', *Lund studies in Geography, Series C, General and Mathematical Geography*, No 1 (Lund, 1966)
4 Op cit, 14
5 Berry, B. J. L., 'Grouping and regionalizing: an approach to the problem of using multivariate methods', *Quantitative geography* (Northwestern Univ. Studies in Geography, No 13 (1967), 219)
6 White, G. F., 'The contribution of geographical analysis to river basin development', *Geographical Journal*, 129 (1963), 421
7 Baker, O. E., 'Agricultural regions of North America', *Economic Geography*, 2 (1926), 459–93
8 Buchanan, R. O., 'Some reflections on agricultural geography', *Geography*, 44 (1959), 1–13
9 Adeemy, M. S., 'Types of farming in North Wales', *Journal of Agricultural Economics*, 19 (1968), 301–16
10 Bennett-Jones, R., 'Farm classification in Britain: an appraisal', *Journal of Agricultural Economics*, 12 (1956), 201–15
11 Rao, C. R., 'The utilization of multiple measurements in

problems of biological classification', *Journal of the Royal Statistical Society*, 10 (1948), 159–203
Rao, C. R., *Advanced statistical methods in biometric research* (Wiley, 1952)

12 King, L. J., *Statistical analysis in geography* (Prentice-Hall, 1969)

13 Berry, B. J. L., 'A method for defining multi-factor uniform regions', *Przeglad Geograficzny* (1961), 263–82
Tarrant, J. R., 'A note concerning the definition of groups of settlements for a central place hierarchy', *Economic Geography*, 44 (1968), 144–51

14 Williams, W. T., and Lambert, J. M., 'Association analysis in plant communities', *Journal of Ecology*, 47 (1959), 83–101
Williams, W. T., and Lambert, J. M., 'Multivariate methods in plant ecology', *Journal of Ecology*, 49 (1961), 717–29
Tarrant, J. R., 'A classification of shop types', *Professional Geographer*, 19 (1967), 179–83

15 Wishart, D., 'Fortran II programs for 8 methods of cluster analysis', *Univ. of Kansas, State geological survey, Computer Contribution*, 38 (1969)

16 Chisholm, M., 'Problems in the classification and use of farming-type regions', *Transactions of the Institute of British Geographers*, 35 (1964), 91–103

17 Tarrant, J. R., 'A note concerning the definition of groups of settlements for a central place hierarchy', *Economic Geography*, 44 (1968), 144–51

18 Chisholm, M., op cit

19 Whittlesey, D., op cit

20 Symons, L., *Agricultural geography* (Bell, 1968)

21 Birch, J. W., 'Observations on the delimitation of farming-type regions: with special reference to the Isle of Man', *Transactions of the Institute of British Geographers*, 20 (1954), 141–58

22 Christian, C. S., and Stewart, G. A., 'General report on the survey of Katherine-Darwin region', *Australian Land Resources Series*, No 1 (CSIRO, 1953)
Perry, R. A. (ed), 'General report on the lands of the Alice Springs area, Northern Territory', *Australian Land Resources Series*, No 6 (CSIRO, 1962)

23 See especially Perry, R. A., op cit

24 Stamp, L. D., 'Fertility, productivity and classification of

land in Britain', *Geographical Journal*, 96 (1940), 389–412

25 *Report of the committee on land utilization in rural areas* (HMSO, Cmd 6378, 1942)

26 Kellogg, C. E., 'Soil and land classification', *Journal of Farm Economics*, 33 (1951), 499–513

27 Hills, G. A., and Portelance, R., *A multiple land-use plan for the Glackmeyer development area* (Ontario dept of lands and forests, 1960)

28 Symons, L., op cit

29 Hilton, N., 'An approach to agricultural land classification in Great Britain', *Land use and resources: studies in applied geography* (Institute of British Geographers, Special Publication, No 1 (1968), 127–42)

30 O'Connor, J., 'Practical problems in classifying land for horticulture', in *Classification of agricultural land in Britain* (Agricultural Land Service, Technical Report No 8, 1962)

31 Hilton, N., op cit

32 Conklin, H. E., 'The Cornell system of land classification', *Journal of Agricultural Economics*, 41 (1959), 548–57

33 Hudson, G. D., 'The unit area method of land classification', *Annals of the Association of American Geographers*, 26 (1936), 99–112

34 Hilton, N., op cit

35 (*a*) Weaver, J. C., 'Crop combination regions in the middle west', *Geographical Review*, 44 (1954), 175–200

(*b*) Weaver, J. C., 'Crop combination regions for 1919 and 1929 in the middle west', *Geographical Review*, 44 (1954), 560–72

(*c*) Weaver, J. C., 'Isotope and compound, a framework for agricultural geography', *Annals of the Association of American Geographers*, 44 (1954), 286–8

36 Buchanan, R. O., 'Some reflections on agricultural geography', *Geography*, 44 (1959), 1–13

37 Weaver, J. C., op cit (a)

38 Thomas, D., *Agriculture in Wales during the Napoleonic wars: a study in the geographical interpretation of historical sources* (University of Wales Press, 1963)

39 Weaver, J. C., op cit (a, b)

40 Scott, P., 'Agricultural regions of Tasmania', *Economic Geography*, 33 (1957), 109–21

41 (*a*) Coppock, J. T., *An agricultural atlas of England and Wales* (Faber, 1964)
(*b*) Coppock, J. T., 'Crop, livestock and enterprise combinations in England and Wales', *Economic Geography*, 40 (1964), 65–81

42 Weaver, J. C., Havg, L. P., and Fenton, B. L., 'Livestock units and combination regions in the mid-west', *Economic Geography*, 32 (1956), 237–59

43 Chisholm, M., op cit

44 Ibid

45 *World Atlas of Agriculture* (Istituto Geografico de Agostini, Novara, 1969)

46 Hartshorne, R., and Dicken, P., 'A classification of agricultural regions of Europe and North America on a uniform statistical basis', *Annals of the Association of American Geographers*, 25 (1935), 99–120

47 Adeemy, M. S., op cit
Birch, J. W., op cit
Church, B. M., *et al*, 'A type of farming map based on agricultural census returns', *Outlook on Agriculture*, 5 (1968), 191–6

48 Whittlesey, D., 'Major agricultural regions of the earth', *Annals of the Association of American Geographers*, 26 (1936), 199–240

49 Buchanan, R. O., op cit

50 Symons, L., op cit

51 Helburn, N., 'The bases for a classification of world agriculture', *Professional Geographer*, 9 (1957), 2–7

52 See Buchanan, R. O., op cit

53 Ibid

54 Maxton, J. P. (ed), *Regional types of British agriculture* (Allen & Unwin, 1936)

55 Adeemy, M. S., op cit

56 Coppock, J. T., op cit (b)

57 *Assistance to small farmers* (HMSO, Cmd 553, 1958)

58 Adeemy, M. S., op cit

59 Ibid

60 Coppock, J. T., op cit (b)

61 Belshaw, D. G. R., and Jackson, B. G., 'Type of farming areas: the application of sampling methods', *Transactions of the Institute of British Geographers*, 38 (1966), 89–93

62 Bennett-Jones, R., *The pattern of farming in the east Midlands* (School of Agriculture, Univ. of Nottingham, 1954)
63 Scott, P., op cit
64 Birch, J. W., op cit, 153
65 Napolitan, L., and Brown, C. J., 'A type of farming classification of agricultural holdings in England and Wales to enterprise patterns', *Journal of Agricultural Economics*, 15 (1962), 595–616
66 *Farm classification in England and Wales* (HMSO, 1967)
67 Church, B. M., *et al*, op cit
68 Adeemy, M. S., op cit
69 Bennett-Jones, R., op cit
70 Ministry of Agriculture, Fisheries and Food and the land utilisation survey of Britain, *Types of farming map of England and Wales* (Geographical Publications Ltd, 1941)
71 Coppock, J. T., op cit (a, b)
72 Symons, L., op cit

CHAPTER FIVE

1 Grigg, D. B., 'Regions, models and classes', in Chorley, R. J., and Haggett, P. (eds), *Models in geography* (Methuen, 1967)
2 Harvey, D., *Explanation in geography* (Arnold, 1969), 221
3 Jones, T. A., 'Soil classification; a destructive criticism', *Journal of Soil Science*, 10 (1959), 196–200
4 Soil survey staff, *Soil classification: a comprehensive system, 7th approximation* (US Dept of Agriculture, 1960)
5 Bourne, R., 'Regional survey and its relation to stock-taking of the agricultural resources of the British empire', *Oxford Forestry Memoirs*, 13 (1931), 16–18
6 Beckett, P. H. T., and Webster, R., *A classification system for terrain. Interim report* (Military Engineering Experimental Establishment, 1965)
7 Prokayev, V. I., 'The facies as the basic and smallest unit of landscape', *Soviet Geography*, 3 (1962), 21–9
8 Anderson, D. J., 'Classification and ordination in vegetation science: controversy over a non-existent problem', *Journal of Ecology*, 53 (1965), 521–6
9 Lambert, J. M., and Dale, M. B., 'The use of statistics in

phytosociology', *Advances in Ecological Research*, 2 (1964), 59–99

10 Zobler, L., 'Statistical testing of regional boundaries', *Annals of the Association of American Geographers*, 47 (1957), 83–95

11 Mackay, J. R., 'Chi square as a tool for regional studies', *Annals of the Association of American Geographers*, 48 (1958), 164

12 Zobler, L., 'Decision making in regional construction', *Annals of the Association of American Geographers*, 48 (1958), 140–8

13 Baker, O. E., 'Agricultural regions in North America', *Economic Geography*, 2 (1926), 459–93

14 Gibson, L. E., 'Characteristics of a regional margin of corn and dairy belts', *Annals of the Association of American Geographers*, 38 (1948), 244–70

15 McCarty, H. H., 'The theoretical nature of land-use regions', *Annals of the Association of American Geographers*, 36 (1946), 97–8

16 Kimble, G. H. T., 'The inadequacy of the regional concept', in Stamp, L. D., and Wooldridge, S. W. (eds), *London essays in geography* (Longmans, 1951)

17 Ellison, W., *Marginal land in Britain* (Bles, 1953)

18 Gibson, L. E., op cit

19 Zobler, L., op cit (1957 and 1958)

20 Christaller, W., *Die zentralen orte in Suddeutschland* (Fischer Verlag, 1933). Translated as Baskin, C. W., *Central places in Southern Germany* (Prentice-Hall, 1966)

21 Berry, B. J. L., and Garrison, W. L., 'Functional bases of the central place hierarchy', *Economic Geography*, 34 (1958), 145–54

22 Vining, R., 'A description of certain spatial aspects of an economic system', *Economic Development and Cultural Change*, 3 (1955), 147–95

23 Jones, W. D., 'An isopleth map of land under crops', *Geographical Review*, 19 (1929)
Jones, W. D., 'Ratio and isopleth maps in the regional investigation of agricultural land occupance', *Annals of the Association of American Geographers*, 20 (1930), 177–95

24 Bunge, W., 'Theoretical geography', *Lund Studies in Geography, Series C, General and Mathematical Geography*, No 1 (Lund, 1966)

25 Sviatlovsky, E. E., and Eells, W. C., 'The centrographic method and regional analysis', *Geographical Review*, 27 (1937), 240–54
Brachi, R., 'Standard distance measures and related methods for spatial analysis', *Papers and Proceedings of the Regional Science Association*, 10 (1963), 83–132
Cole, J. P., and King, C. A. M., *Quantitative geography* (Wiley, 1968)
King, L. J., *Statistical analysis in geography* (Prentice-Hall, 1969)
26 Tarrant, J. R., 'Some spatial variations in Irish agriculture', *Tijdschrift voor Economische en Sociale Geografie*, 60 (1969), 228–37
27 Blumenstock, D. I., 'The reliability factor in drawing isarithms', *Annals of the Association of American Geographers*, 43 (1953), 289–304
28 Robinson, A. H., and Sale, R. D., *Elements of cartography* (3rd ed, Wiley, 1969)
29 Harbaugh, J. W., and Merriam, D. F., *Computer applications in stratigraphic analysis* (Wiley, 1968)
30 Mackay, J. R., 'Some problems and techniques in isopleth mapping', *Economic Geography*, 27 (1951), 1–9
Mackay, J. R., 'The alternative choice in isopleth interpolation', *Professional Geographer*, 5 (1953), 2–4
31 Robinson and Sale, op cit
32 Mackay, J. R., 'An analysis of isopleth and choropleth class intervals', *Economic Geography*, 31 (1955), 71–81
Mackay, J. R., 'Isopleth class intervals: a consideration of their selection', *Canadian Geographer*, 7 (1963), 42–5
33 Wright, J. K., 'Map makers are human: comments on the subjectivity in maps', *Geographical Review*, 32 (1942), 527–44
34 Mackay, J. R., op cit
35 Harbaugh and Merriam, op cit
36 Cole, A. J., 'An iterative approach to the fitting of trend surfaces', *Kansas state geological survey*, *Computer contribution*, No 37 (Univ. of Kansas, 1969)
37 Rosing, K. E., 'Computer graphics', *Area*, 1 (1969), 2–7
38 Cole and King, op cit
39 Tarrant, J. R., op cit
40 Harbaugh and Merriam, op cit
41 Chorley, R. J., and Haggett, P., 'Trend surface mapping in

geographical research', *Transactions of the Institute of British Geographers*, 37 (1965), 47–67
42 Gittins, R., 'Trend surface analysis of ecological data', *Journal of Ecology*, 56 (1969), 845–69
43 Coppock, J. T., and Gillmor, D. A., 'The cattle trade between Ireland and Great Britain', *Irish Geography*, 5 (1967), 320–6
44 Yamane, T., *Statistics: an introductory analysis* (Harper, 1967), Chapter 14
45 Freeman, T. W., *Ireland: a general and regional geography* (4th ed, Methuen, 1969), 184
46 Allen, P., and Krumbein, W. C., 'Secondary trend components in the top Ashdown pebble beds: a case history', *Journal of Geology*, 70 (1962), 507–38
47 Norcliffe, G. B., 'On the use and limitations of trend surface models', *Canadian Geographer*, 13 (1969), 338–48
Tarrant, J. R., 'Comments on the use of trend surface analysis in the study of erosion surfaces', *Transactions of the Institute of British Geographers*, 51 (1970), 221–2
Unwin, D. L., and Lewin, J., 'Some problems in the trend analysis of erosion surfaces', *Area*, 3 (1971), 13–14
48 Tarrant, J. R., op cit (1970)

CHAPTER SIX

1 Harvey, D., *Explanation in geography* (Arnold, 1969)
2 Burton, I., 'The quantitative revolution and theoretical geography', *Canadian Geographer*, 7 (1963), 156
3 Berry, B. J. L., 'Further comments concerning "Geographic and Economic" economic geography', *Professional Geographer*, 11 (1959), 12
4 Bunge, W., 'Theoretical geography', *Lund Studies in Geography, Series C, General and Mathematical Geography* (Lund, 1966)
5 Hartshorne, R., *The nature of geography: a critical survey of current thought* (Association of American Geographers, 1939)
6 Lukerman, F., 'Towards a more geographic economic geography', *Professional Geographer*, 10 (1958), 2–10
7 Robinson, A. H., Lindberg, J. B., and Brinckman, L. W., 'A correlation and regression analysis applied to rural farm popu-

lation densities in the Great Plains', *Annals of the Association of American Geographers*, 51 (1961), 211

8 Manley, V. P., and Olmstead, C. W., 'Geographical pattern of labour input as related to output indices of scale of operation in American agriculture', *Annals of the Association of American Geographers*, 55 (1965), 629–30

9 Harvey, D., op cit

10 Bunge, W., op cit, 6

11 Jones, E., 'Cause and effect in human geography', *Annals of the Association of American Geographers*, 46 (1956), 369–77

12 Op cit, 367

13 Hildore, J. J., 'The relations between cash-grain farming and landforms', *Economic Geography*, 39 (1963), 84–9

14 Yamane, T., *Statistics: an introductory analysis* (Harper, 1967), 436

15 Siegal, S., *Non-parametric statistics for the behavioral sciences* (McGraw-Hill, 1956)

16 Tarrant, J. R., 'A classification of shop types', *Professional Geographer*, 19 (1967), 179–83

17 King, L. J., *Statistical analysis in geography* (Prentice-Hall, 1969)

18 Robinson, A. H., *et al*, op cit

19 Child, D., *The essentials of factor analysis* (Holt, Rinehart and Winston, 1970)

20 King, L. J., op cit

21 Cattell, R. B., *Factor analysis* (Harper, 1952)

22 Tarrant, J. R., *Computers in the environmental sciences* (Geoabstracts, 1972)

23 Cattell, R. B., *Handbook of multivariate psychology* (Rand McNally, 1966)

24 Berry, B. J. L., 'A method for devising multi-factor uniform regions', *Przeglad Geograficzny*, 33 (1961), 263–79

25 Burt, C., and Banks, C., 'A factor analysis of body measurements for British adult males', *Annals of Eugenics*, 13 (1947), 238–56

26 Vernon, P. E., *The structure of human abilities* (Methuen, 1950), 130

27 Harman, H. H., *Modern factor analysis* (Univ. of Chicago Press, 1967)

28 Henshall, J., 'The demographic factor in the structure of

Barbados agriculture', *Transactions of the Institute of British Geographers*, 38 (1966), 183–96

29 Munton, R. J. C., and Norris, J. M., 'The analysis of farm organisation: an approach to the classification of agricultural land in Britain', *Geografiska Annaler, Series B*, 52 (1969), 95–103

30 Ibid

CHAPTER SEVEN

1 Smith, E. D., 'Voluntary group programs to meet farmer marketing problems', in Dubov, I. (ed), *Contemporary Agricultural Marketing* (Univ. of Tennessee Press, 1968)

2 Kohls, R. L., *Marketing of agricultural products* (Macmillan, 1967)

3 Ibid

4 Ministry of Agriculture, *The working of the Agricultural Marketing Acts* (HMSO, 1947)

5 United States Department of Agriculture, 'The conservation reserve program of the soil bank', *Agriculture Information Bulletin*, No 185 (1958)

6 Shitze, R. G. F., 'Policy direction and economic interpretations of the US Agricultural Act of 1970', *Journal of Agricultural Economics*, 23 (1972), 99–108

7 United States Department of Agriculture, 'The effects of acreage-allotment programs 1954 and 1955. A summary report', *Production Research Report*, 3 (1956)

8 Warley, T. K. (ed), *Agricultural producers and their markets* (Blackwell, 1967)

9 *Report of the committee of inquiry into fatstock and carcase meat marketing and distribution* (HMSO, Cmd 2282, 1964)

10 *Horticultural marketing. Report of the committee* (HMSO, Cmd 61, 1957)

11 Britton, D. K., *Cereal marketing in Britain* (Pergamon, 1969)

12 Hazeldine, W. J., 'Cereals—forward contracting', *Agriculture*, 74 (1967), 314–16

13 Warley, T. K., op cit

14 Ministry of Agriculture, op cit

15 Strauss, E., and Churcher, E. H., 'The regional analysis of the milk market', *Journal of Agricultural Economics*, 18 (1967), 221–40

16 Kohls, R. L., op cit
17 Silvey, D. K., 'Cereals in the common market', *Agriculture*, 74 (1967), 230–4
18 Clout, H., *Agriculture: studies in contemporary Europe* (Macmillan, 1971)
19 Britton, D. K., op cit
20 Colman, D. R., *The United Kingdom cereal market: an econometric investigation into the effects of pricing policies* (Manchester Univ. Press, 1972)
21 Hewlett, R., 'Status, achievements and problems of agricultural co-operatives in Europe', in Warley, T. K., op cit
22 Kohls, R. L., op cit
23 Agricultural adjustment unit, 'Farm size adjustment', *Bulletin No. 6* (Univ. of Newcastle, 1968)
24 Bucksar, R. G., 'Significant changes in the American egg industry', *Journal of Geography*, 67 (1968), 36–41
25 Guiton, N. F. le H., 'Contracts in agricultural marketing', in Warley, T. K., op cit
26 Allen, G. R., 'An appraisal of contract farming', *Journal of Agricultural Economics*, 33 (1972), 89–98
27 Watts, H. D., 'The location of the beet-sugar industry in England and Wales, 1912–36', *Transactions of the Institute of British Geographers*, 53 (1971), 95–116
28 Dalton, R. T., 'Peas for freezing: a recent development in Lincolnshire agriculture', *East Midlands Geographer*, 5 (1971), 133–41
29 Ibid
30 Coppock, J. T., *An agricultural geography of Great Britain* (Bell, 1971)

CHAPTER EIGHT

1 Edwards, A. M., and Wibberley, G. P., 'An agricultural land budget for Britain 1965–2000', *Wye College, Studies in Rural Land Use*, No 10 (1971)
2 Gregor, H. F., *Geography of agriculture: themes in research* (Prentice-Hall, 1970)
3 Best, R. H., 'Competition for land between rural and urban uses', Chapter 6 in *Land-use and resources: studies in applied geography* (Institute of British Geographers, Special Publication No 1 (1968), 89–100)

Best, R. H., 'The extent of urban growth and agricultural displacement in post-war Britain', *Urban Studies*, 5 (1968), 1–23

Bogue, D., 'Metropolitan growth and the conversion of land to non-agricultural uses', *Studies in Population Distribution*, No 11 (Oxford, Ohio, 1956)

Wibberley, G. P., 'Land scarcity in Britain', *Journal of the Town Planning Institute*, 53 (1967), 129–36

4 Edwards and Wibberley, op cit

5 Fogarty, F., 'Land: a new kind of boom', *Architectural Forum*, 98 (1957), 101–6 and 230–2

6 Gillies, J., and Mittelbach, F., 'Urban pressures on California land: a comment', *Land Economics*, 34 (1958), 80–3

Clawson, M., *Suburban land conversion in the United States: an economic and governmental process* (Resources for the future, 1971)

7 Hoyt, H., 'The growth of cities from 1800 to 1960 and forecasts to the year 2000', *Land Economics*, 39 (1963), 167–74

8 Clawson, M., op cit

9 Break, G., 'The sensitivity of housing demands to changes in mortgage credit terms', in *The economic impact of Federal loan insurance* (Washington DC national planning association, 1961), 225–48

10 Gregor, H. F., 'Urban pressures on California land', *Land Economics*, 33 (1957), 311–25

11 Muth, R. F., 'Economic change and rural-urban land conversions', *Econometrica*, 29 (1961), 1–23

12 Blase, M. G., and Staub, W. J., 'Real property taxes in the rural-urban fringe', *Land Economics*, 47 (1971), 168–74

13 Griffin, P. F., and Chatham, R. L., 'Urban impact on agriculture in Santa Clara Co., California', *Annals of the Association of American Geographers*, 48 (1958), 195–208

14 Gregor, H. F., op cit

15 Griffin and Chatham, op cit

16 Gregor, H. F., op cit

17 Clawson, M., op cit

18 Northam, R. M., 'Vacant urban land in the American city', *Land Economics*, 47 (1971), 345–55

19 Wehrwein, G. S., 'The rural-urban fringe', *Economic Geography*, 18 (1942), 217–28

20 Glaab, C. N., and Brown, A. T., *A history of urban America* (Macmillan, 1967)
21 Boyce, R. R., 'The edge of the metropolis: the wave theory analog approach', *British Columbia Geographical Series*, 7 (1966), 31–40
22 Pryor, R. J., 'Defining the rural-urban fringe', *Social Forces*, 47 (1968), 202–15
23 Griffin and Chatham, op cit
Gregor, H., op cit
24 Gillies and Mittelbach, op cit
25 Lessinger, J., 'The case for scatteration: some reflections on the national capital regional plan for the year 2000', *Journal of the American Institute of Planners*, 28 (1962), 159–69
26 Clawson, M., op cit
27 Edwards and Wibberley, op cit

ACKNOWLEDGEMENTS

This book could not have been written without the constant advice and encouragement given to me by many of my friends and colleagues. Such assistance has been given too often to allow individual mention but I would like to take this opportunity to make a general acknowledgement. Special thanks go to Sue Middlege who typed the manuscript and to Barbara Satchell and David Mew who drew the maps.

Figures 3, 4, 7, 8, 9, 10, 12, 16, 17, 18, 22, 29, 39, 40 and 42 are reproduced by kind permission of the authors and publishers as indicated in the captions.

Finally, and by no means least, I would like to thank the series editor Professor R. Lawton for his patient reading and correction of my original text. Any inadequacies, errors or omissions remain entirely my responsibility.

JOHN TARRANT

Norwich
February 1973

INDEX

H

I

J

K

L